Science, politics and society in early nineteenth-century Ireland

MANCHESTER 1824

Manchester University Press

Science, politics and society in early nineteenth-century Ireland

The Reverend William Richardson

Allan Blackstock

Manchester University Press

Manchester and New York

distributed in the United States exclusively
by Palgrave Macmillan

Published by Manchester University Press
Oxford Road, Manchester M13 9NR, UK
and Room 400, 175 Fifth Avenue, New York, NY 10010, USA
www.manchesteruniversitypress.co.uk

Distributed in the United States exclusively by
Palgrave Macmillan, 175 Fifth Avenue, New York,
NY 10010, USA

Distributed in Canada exclusively by
UBC Press, University of British Columbia, 2029 West Mall,
Vancouver, BC, Canada V6T 1Z2

British Library Cataloguing-in-Publication Data
A catalogue record for this book is available from the British Library

Library of Congress Cataloging-in-Publication Data applied for

ISBN 978 0 7190 8518 5 *hardback*

First published 2013

Typeset in Caslon by R. J. Footring Ltd
Printed in Great Britain
by MPG Books Group, UK

For Alice, the artist in the family

BELFAST NATURAL HISTORY & PHILOSOPHICAL SOCIETY

QUID RETRIBUAMUS PRO TANTO

1821

Contents

Preface

William Richardson (1740–1820) died in interesting times. His death is recorded as occurring on Wednesday 14 June 1820 at his Clonfeacle rectory in county Tyrone. The *Belfast News-Letter*, which carried his obituary, was also filled with news items which would have interested this idiosyncratic cleric, including reports of London's Queen Caroline riots and of the funeral of the great Irish patriot Henry Grattan. Economic concerns featured prominently. The paper highlighted bank failures throughout Ireland but noted that landed support had stopped a run on the Belfast banks and restored confidence. Reports on parliamentary debates about protective duties on Irish goods noted that Irish MPs, who favoured continuance of this Union arrangement, disagreed with the free trader David Ricardo. Readers complained of the English poor law returning Irish paupers to an increasingly overcrowded island. Richardson's obituary was followed by that of Sir Joseph Banks, President of the Royal Society, which noted how on Captain Cook's 1767 voyage Banks had used the Linnaean system to classify new plants. If Banks's botany facilitated imperialism in the South Seas, technological advance complemented the Union's consolidating logic. A steam packet named after Walter Scott's hero, Rob Roy, sailed from Belfast to Glasgow three times a week.[1] But Richardson, an Irish provincial amateur whose geological and botanical ideas penetrated the highest metropolitan scientific circles, lived in even more interesting times.

Richardson had experienced the horrors of the 1798 rebellion, personally encountered members of the first Orange lodges and reacted to demands for Catholic emancipation. Intellectually, he participated in the development of provincial scientific culture in Ireland and lived through the emergence of disciplinary specialisation in geology and botany. His friends included brilliant romantics like Humphry Davy and Enlightened connoisseurs such as Thomas Percy, Bishop of Dromore. His correspondence network crossed political and national boundaries, embracing English and Scottish and Irish noblemen, European scientists and American farmers. Because Richardson's interests often overlapped, this book does not follow the chronological pattern of a classical biography. But, as befits a polymath, it adopts a thematic structure dictated by his interests in politics, geology, agricultural botany and political economy.

A prolific writer in the pamphlet and periodical literature, Richardson generated, sometimes at length, an extensive private correspondence. However, no personal archive survives and the oeuvre of his work needed to be reconstructed from, to misquote his own evocative phrase, the 'scattered remnants' of what must once have been a vast collection. A few historians have tentatively dipped into certain of his works as an adjunct to their particular interests; but an integrated, three-dimensional picture of this significant figure has never yet been attempted. This has now been made possible by drawing together the extensive surviving printed material from periodicals and pamphlets located in Ireland, England, Scotland and Denmark and unearthing letters from the archives of Richardson's correspondents. Much of this material is unused, and some of it has only recently entered the public domain, particularly the extensive correspondence – over 100 letters – from Richardson to George Bellas Greenough, Member of Parliament and first President of the Geological Society. This collection is the result of a recent amalgamation of Greenough's papers from the University of Cambridge and University College London. It is vital in understanding Richardson's work as it forms a private commentary on his public material and contains eyewitness accounts of popular political movements like the Orange Order, Defenders and Freemasons in Ulster.

This marriage of printed and manuscript material forms the archival core of the book, which does not simply reconstruct the story, however interesting, from these sources. Rather, it seeks to position the specific analysis of Richardson, as far as possible, in the wider historical literature relevant to whichever theme is under consideration. This analysis is informed by recent trends in the historiography of science to identify the cultural and associational basis for the construction, dissemination and reception of knowledge and its interactions with politics and religion.[2] In this respect the book follows the stance outlined in Roy Porter's study[3] of the development of geology as a scientific discipline insofar as the eventual dismissal of Richardson's ideas is less important than his contribution to contemporary debate. This counter-anachronistic position is broadened to embrace botany as well as geology: put simply, the construction of ideas and polemics in advocacy or defence of them cannot be divorced from their contemporary context.

As well as outlining and contextualising these ideas, this book also seeks to look beyond Richardson the man to see what wider patterns he represents. It is argued that Richardson occupied an intermediary role in both politics and science, allowing an assessment of his historical significance in each area. This can be seen from his earlier life as a college don and, particularly, as a rural Ulster rector in the 1780s and 1790s. The period immediately preceding the 1798 rebellion saw the emergence and growth of popular plebeian movements which engaged in protest and sectarian feuding. Richardson's clerical office gave him a key place in rural society, positioned between the county elite and the lower orders, but linked to both. In Chapter 2, the identification and analysis of his political activities permit a more nuanced view of Ulster politics at this period, covering popular movements. Furthermore, after the Act of Union Richardson's links with Greenough, originally established for scientific reasons, functioned as a conduit to promote and justify Protestant Ascendancy in the unfamiliar and uncomprehending environment of Westminster. In this sense Richardson anticipated later parliamentary hostility to the dominant Irish variety of loyalism which culminated in the 1835 Select Committee reports on Orangeism. Chapter 3 investigates Richardson's geological activities, which first brought him to the notice of the scientific community nationally and internationally.

Though Richardson intervened in scientific and intellectual debates, the role played by his personality in conditioning his reactions is investigated. A character emerges whose contradictions clashed with his virtues. Richardson was highly intelligent but naive, breathtakingly learned in some respects and ignorant in others. He could be snobbish and tolerant, politically independent yet deferential, one minute self-deprecating, bombastic the next. Stubborn, stoic and utterly fearless, he was a man conditioned by the eighteenth century who flourished in the early nineteenth. The book explores the question of why Richardson's activity as a writer began so late in life, when he was over sixty; yet, like his 'protégé' fiorin grass, grew luxuriantly in a relatively short space of time. What details remain from his earlier life at Trinity College Dublin and of his day-to-day life as rector of Clonfeacle are garnered to explain his long gestation as a public man. A further theme to be investigated is Richardson's role as a provincial Protestant clergyman and he will be compared with others of a similar ilk. He will also be compared with the best-known Protestant publicist of all, Sir Richard Musgrave. Although Richardson did not publish until 1801, as the book's later chapters show, when he did emerge in public he was hard to ignore.

Notes

1 *Belfast News-Letter*, 13, 16, 20, 23, 27 June 1820.
2 M. Fissell and R. Cooter, 'Exploring natural knowledge: science and the popular', in R. Porter (ed.), *The Cambridge history of science, vol. 4: eighteenth-century science* (Cambridge, 2003), pp. 131–5.
3 R. Porter, *The making of geology: earth science in Britain, 1660–1815* (Cambridge, 1977).

Acknowledgements

Special thanks are due to Dr Eoin Magennis for suggesting the project and for support during writing, to the University of Ulster's Arts and Humanities Research Institute for study leave and to the Belfast Natural History and Philosophical Society for assistance with the cover, designed by George Meharg. The following individuals have helped with advice: Dr Frank Ferguson, Professor David Hayton; Dr Sophie Hillan; Dr Andrew Holmes; the late Professor Peter Jupp; Dr A. P. W. Malcomson; Trevor Parkhill; the Rev. Nigel Playfair; Dr William Roulston; and Dr Andrew Sneddon. Thanks are also due to Dr Petri Mirla and Catherine Gartland for translations from Danish and French, and to Stephen Kirkwood for Latin translations. I am also grateful to Mrs Sarah Landale of Dalswinton Estate, Dumfriesshire, for allowing me to view Clonfeacle Tower, erected in 1810 to honour William Richardson.

For permission to publish, my thanks are due to: the Deputy Keeper of the Records (PRONI); Dr William Drennan; the Trustees of the British Library; the National Library of Ireland; Trinity College Dublin; the National Archives of Ireland; the National Archives of Scotland; the National Library of Scotland; the National Trust; the Lennoxlove Charitable Trust and the Duke of Hamilton; the Royal Institution; University College London; the University of Glasgow; Armagh Diocesan Office; the Ulster Museum; the Linenhall Library; the Royal Dublin Society; the Robinson Library; the O'Feigh Library; Kent Archives Office; Suffolk County Record Office; Hants County Record Office. The Home Office 100 Series are Crown copyright.

Abbreviations

Periodicals

BJHS	*British Journal of the History of Science*
ECI	*Eighteenth-Century Ireland*
EHR	*English Historical Review*
HJ	*Historical Journal*
IHS	*Irish Historical Studies*

Institutions

BL	British Library
HO	Home Office
LHL	Linenhall Library, Belfast
NAI	National Archives of Ireland
NAS	National Archives of Scotland
NLI	National Library of Ireland
NLS	National Library of Scotland
PRONI	Public Record Office of Northern Ireland
UCL	University College London

1

'Virtue appears like an Oak':[1] William Richardson's family and background

This motto from the Richardson family crest is certainly appropriate, for William Richardson saw himself as a virtuous man. Yet everyone who knew him found a tenaciously, often belligerently, stubborn man. He conformed to the general eighteenth-century conception of the patriotic, public-spirited citizen being a virtuous man; but Richardson made a virtue from the necessity always to be right. This self-righteous trait revealed itself in a very strict sense of propriety. Richardson reacted to contradiction with explosive anger, sullen sulkiness or scathing sarcasm. He was not a man to cross lightly. Even his own family treated him cautiously. As a prominent pamphleteer, Richardson was a public character, but personality is vital in the formation of any such character. Richardson once admitted that 'I love controversy'.[2] This opening chapter considers formative influences on Richardson's personality, such as ancestry, upbringing and education, and reviews his actions before he stormed into the world of print. It argues that the pugnacious polemicist of the early nineteenth century was fully formed in the eighteenth.

The Richardson family

The Richardson family was solidly Protestant and deeply rooted in the Ulster soil. The seventeenth-century Plantation, which heralded major changes in landholding, indirectly brought William Richardson's ancestors to Ireland. Plantation grants of 1,000 acres in the Precinct of Mountjoy, county Tyrone, were made to Bernard and Robert Lindsay from Haddington in Scotland in 1610. By 1618 these land grants had been transferred to another Scot who came from the same area, Alexander Richardson. Pynnar's 1618 report on the Ulster Plantation noted that these lands contained a fortified stone bawn, a timber house and sufficient tenants to produce '39 men with arms'.[3] Several of Richardson's ancestors fought at the siege of Derry and were proclaimed traitors by the 'Patriot Parliament' in 1689.[4] After the Jacobite defeat at the Boyne in 1690 had negated any land forfeiture made under the Act of Attainder, an Archibald Richardson sat for the Tyrone borough of Augher in the 1692 Williamite parliament. His brother, William, held the seat between 1737 and his death in 1755, when it passed to his son, St George Richardson, a professional soldier.[5] The origins of the family's county Londonderry connections are obscure, but strong links were developed with the Irish Society, comprised of London companies which had received Plantation grants. One source has Archibald's brother William as managing agent for the Merchant Taylor's Company estate near Coleraine. He purchased this estate in 1729 and became the Irish Society's general agent for county Londonderry. He also leased the lucrative Bann fishery from the Irish Society in 1724 and appointed his neighbour, Hercules Heyland, to be chief fishing agent.[6] These ancestral themes found their echo in the Reverend William Richardson, who, like his friend Humphry Davy, was a keen angler, and also he devised schemes to make the River Bann navigable. Richardson also inherited the family's Protestantism, though more as a political cause than as a doctrinal position. But this was not the only legacy.

The Richardsons had longstanding clerical and literary traditions. William Richardson, the MP for Augher, was a friend and correspondent of Dean Swift.[7] His brother, our William Richardson's grandfather, John, was rector of Belturbet, county Cavan, and dean of Kilmacduagh. He also knew Swift and was a strong Protestant who had unsuccessfully tried to convert 'the Popish natives' by preaching in Gaelic. He travelled to London in 1711 to get backing for Irish translations of the New Testament, a catechism and Book of Common Prayer for parish schools to promote Protestantism.[8] This enterprise bankrupted him and forced the sale of family lands at Orator, county Tyrone, worth £3,000 a year.[9] Two of John Richardson's sons, John and James, were also Trinity-educated clerics with livings in Tyrone and Londonderry. Another son, Charles Richardson, was married to Sarah or 'Sally', the daughter of Hercules Heyland of Castleroe. In 1740, they produced their only child, christened William after his grand-uncle, who would become the rector of Clonfeacle.[10]

Sally Richardson was widowed young in 1743. According to family legend, she was 'very little, very pretty' and had a noticeably vivacious personality, which earned her the revealing nickname of 'the pocket Venus'. Significantly, Sally 'devoted her life to her only son … the celebrated Dr Richardson', an indulgence which undoubtedly influenced his personality.[11] This anxious maternal solicitude may have been exacerbated by his height. As a child, William Richardson seemed to have inherited his mother's stature, being, in his own words, once considered 'dwarfish', though he continued growing until the age of twenty-three.[12] This early combination of spoiling and sheltering made the young man feel his singularity and importance, and his education enhanced this.

Education

Little is known about Richardson's early education, except that he attended Derry Diocesan School (then known as the Derry Free School) under the Reverend John Torrens. This school was originally endowed in 1617 by Matthew Springham of the Irish Society and had strong links with the Church of Ireland and Trinity College Dublin. Torrens was recommended to his position by Derry Corporation as 'an eminent scholar' whose 'excellent' school 'rose to fame under his management'. Richardson and Torrens had a shared past, which encouraged a close relationship – both had ancestors at the siege of Derry. Richardson certainly received much individual attention, which fed his sense of his own significance. Even under Torrens's successor, the school accommodated only four boarders and sixteen day scholars. Richardson's lifelong love of classical literature possibly reflected Torrens's influence, as Trinity instructed school principals on the recommended texts for prospective entrants.[13] However, it was Richardson's time at university which really moulded his mind and established his status.

In the eighteenth century, Trinity College Dublin was at the pinnacle of the Irish education system and its fellows (mostly clergymen) 'presided over the intellectual and cultural formation' of Anglican clerics and the broader Protestant community.[14] Richardson entered the College in April 1759 and, though older than most junior freshmen, who were in their mid to late teens, he would have appeared younger than his nineteen years due to his small stature. Like most students, he entered as a 'pensioner'. There was a strict social hierarchy, ranging from 'sizars', through 'pensioners', who paid £7.10.0 per half year, to 'fellow commoners', who paid bi-annual fees of £15 and sons of noblemen who paid £30.[15] Pensioners were usually 'sons of persons of moderate income'. Wolfe Tone was a Trinity pensioner whose father was a respectable coachmaker who could afford a substantial house and servants.[16] Richardson's family background was landed; however, with his father long dead, being a pensioner represented his widowed mother's economic position rather than perceptions of rightful social status. The social category difference was tangible and marked by dress and privilege. Fellow commoners and 'noblemen', as Tone enviously remarked, wore a gown even more splendid than a fellow's, being 'as full of tassels as a livery servant's'. Pensioners had their privileges too. They could dine together, wear a fine gown with 'hanging sleeves and tassels', have access to the gallery in the Irish House of Commons and recognition as gentlemen in Dublin. Students could be elevated to 'scholars', as Richardson was in 1761, but the process was exacting. Edmund Burke became a scholar in 1746 only after two days' examination by the fellows in Greek and Roman authors.[17]

Undergraduates were divided into year classes, junior and senior freshmen and junior and senior sophisters, with each group having separate lectures. Junior freshmen traditionally began studying rhetoric and logic and deepened these studies as senior freshmen. Natural science was introduced for junior sophisters, followed by ethics in the fourth year.[18] The courses were originally prescribed in the statutes by Archbishop Laud, but in 1761 were modified to reflect the needs of the time.[19] The curriculum was dominated by John Locke in epistemology and politics and by Isaac Newton and Robert Boyle's experimental philosophy in science.[20] In genuinely Whiggish spirit, the new elements sat comfortably with the old. A new logic text, Richard Murray's *Artis logicae compendium*, was still purely Aristotelian. In Richardson's time the subjects were classical learning, oriental, ancient and modern languages, criticism, history, oratory, logic, ethics and metaphysics, natural and experimental philosophy, anatomy, botany, chemistry, mathematics, civil and canon law, theology and ecclesiastical history.[21] The BA curriculum reflected the enlightened *Zeitgeist*. Senior sophisters studied ethics but read only one book on theology, John Conybeare's *Defence of revealed religion* (1732).[22] Yet change did not come fast enough for everyone. Jonah Barrington (who entered in 1773) described the curriculum as 'learned' but 'ill arranged'. Most entrants were teenagers and Barrington reckoned they studied Locke on human understanding before they were sufficiently mature to have developed their own. He also believed the curriculum favoured 'abstruse sciences' like optics, natural philosophy, astronomy, mathematics and metaphysics without the leavening of *belles-lettres*, history and geography. This dyspeptic view may reflect Barrington's indifferent academic performance.[23] Moreover, subjects not covered by the official curriculum were available from the College Historical Society, formed in 1770 'for the cultivation of historical knowledge and the practice of the members in oratory and composition'.[24] The curriculum privileged the classics, which even Barrington conceded formed 'essential parts of a gentlemen's education'.[25] Students were expected to read over twenty authors, from Homer to Justinius, and to know the entire *Iliad*, *Odyssey* and *Aeneid*.[26] Classics overlapped with science as they were seen as providing an understanding of the natural world. Oliver Goldsmith traced his interest in nature to reading Pliny.[27]

Richardson completed his Bachelor of Arts degree in 1763, an exacting process involving oral examinations in classics and science and the disputing of syllogisms. In 1766 he took his Master of Arts degree and was elected a fellow.[28] The fellowship examination, said his contemporary Patrick Duigenan, was 'the severest and most solemn … on any part of the earth'. Hyperbole aside, the procedure was undoubtedly daunting: four hours of questioning spread over four days, conducted publicly before learned citizens and university dons. The Provost and senior fellows sat opposite the applicant and quizzed him on logic, mathematics, natural and moral philosophy, chronology, history and the learned languages. In reply to this intellectual barrage the doughty applicant had to answer loudly enough for the entire audience to hear.[29] The rewards justified the ordeal, as fellowships were prestigious. The College population had reached 500 by the 1770s but only twenty-two of these were fellows. Fellows could vote for the College's MPs, and though junior fellows earned less than seniors they still received 'considerable' fees from tutoring pupils. Richardson had some distinguished fellows for company, including the top classical scholar Thomas Leland, who was a senior fellow, as was the logician Richard Murray, who became Provost in 1799. Other notable fellows were Hugh Hamilton, who authored a treatise on conic sections and was also a fellow of the Royal Society, and the historian Dr Michael Kearney.[30] Having entered as a pensioner, Richardson appreciated the fellowship's intellectual kudos, always signing his pamphlets 'William Richardson DD Late Fellow of Trinity College Dublin'.[31]

Trinity's Anglican and Protestant ethos imbued College life. Students were expected to attend divine service and could not be elevated to scholars unless they had done so. As a fellowship candidate, Richardson took oaths abjuring 'Pontifical religion' and papal authority.[32] Though there was no formally constituted divinity school, a chair was established in his time and Richardson completed his Bachelor of Divinity degree and in 1778 his divinity doctorate.[33]

Barrington thought the Trinity atmosphere 'pedantic'. Yet one man's pedantry was another's civility. Richardson's College days overlapped with the reign of two Provosts, Francis Andrews and John Hely-Hutchinson, both of whom, in their own way, left their mark. Andrews was a Derryman who never lost his provincial accent, yet epitomised the enlightened *Zeitgeist*, having experienced the Grand Tour, made elevated social contacts and raised Trinity's prestige with chairs established in divinity and Greek. The natural sciences also flourished and Andrews ensured that Trinity's social repute matched its academic standing. In 1761 a London visitor enjoyed 'a most magnificent entertainment' in a room 'elegant and well lighted up with a profusion of wax lights, claret and scarlet gowns'.[34] Two years after Richardson left, this intellectual endeavour found an outlet in the Royal Irish Academy, founded to promote scholarship in science, literature and antiquities. Andrews died in 1774 and posthumously endowed a chair of astronomy, a subject which, with its connection to natural theology and Newtonian philosophy, suited clerical educationalists. John Hely-Hutchinson was politically ambitious and became very unpopular with the fellows for running the College like a pocket borough and interfering in fellowship elections.[35] Barrington said that Hutchinson tried to modernise the curriculum by introducing modern languages, 'to adapt the course to men of rank as well as men of science', but was baulked by 'pedantic fellows'.[36] Being a stickler for protocol, Richardson was probably among the pedants and he later recalled how Hely-Hutchinson contravened the fellows' traditional obligation to celibacy by manipulating marriage dispensations.[37]

In the early eighteenth century, Trinity was believed to harbour Jacobite sympathisers, but the dominant ethos became 'liberal Whiggery' until the political ruptures of the 1790s.[38] In Richardson's time, views ranged from the doctrinaire conservatism of Patrick Duigenan and John Toler, to those, like himself, Whiggishly disposed but tending towards conservatism. There were also more reform-minded Whigs, like Henry Grattan and John Parnell, and radicals, like Wolfe Tone and Whitley Stokes, who became United Irishmen.[39] Williamite commemorations were not considered sectarian until the 1790s and Richardson approved of these displays at Trinity, noting how William III's birthday was 'celebrated under my window in grand procession highly decorated with orange'.[40] The Historical Society represented this broad spectrum of opinion by debating issues like free trade and the American War. In 1782, the year of the Irish parliament's legislative independence, its Members included Tone and Thomas Addis Emmet, as well as the more moderate Laurence Parsons. Though Richardson had left College by the time of Lord Clare's 'visitation' of 1798, when nineteen radicals and revolutionaries were expelled, he would have wholeheartedly supported it.[41] Indeed, he came to detest Grattan's 'hostility to Protestants, Loyalists, Clergy and College'.[42]

College life, as both student and tutor, provided many opportunities for building networks. If Trinity helped shape Richardson politically, it also gave many social and intellectual contacts. Judging by parental occupations, Richardson's fellow students came from a wide social and economic range, including landowning gentry and nobility, merchants, lawyers, bankers, blacksmiths and glaziers. His contemporaries included Lord Gosford's son, Arthur Acheson, and John Toler, later Lord Norbury. Toler was an outspoken judge, anti-Catholic and anti-liberal, who Barrington claimed 'had a hand for every man and a heart for nobody'.[43] Toler tried key United Irishmen, including Robert Emmet, in whose 1803 insurrection Richardson's friend and fellow student Arthur Wolfe, then Lord Kilwarden, was murdered.[44] Richardson's relationship with Wolfe epitomises the Trinity network in operation. As Attorney General, Wolfe's legal peregrinations during sectarian trouble in mid-Ulster renewed the contact. Richardson used Wolfe as a go-between with the government when the yeomanry was formed in 1796. The friendship was so close that Richardson named his second son after him.[45] John Fitzgibbon, Lord Clare, was another ultra-loyalist legal man whose student days overlapped with Richardson's.[46] Richardson also formed acquaintances with scientific men. The inventor and agricultural improver Richard Lovell Edgeworth entered a year after Richardson (though he was soon withdrawn and sent to Oxford).[47] Laurence Parsons entered during Richardson's fellowship. He became a College MP between 1782 and 1783, and in 1807 inherited the title of second Earl of Rosse and the family Parsonstown estate. Though, unlike Richardson, Toler or Wolfe, Parsons was leniently disposed to the 1798 rebels, in intellectual and scientific terms they had much in common. Parsons was a renowned astronomer and antiquarian, as well as being an agricultural improver and a member of the Royal Irish Academy. Richardson exploited their mutual interest in bog improvement, visited Parsonstown and dedicated a pamphlet on the Bog of Allen to him.[48]

Arthur Young noted that Trinity students' subsequent careers were often predetermined by parental profession or family tradition. As well as landowning, the law was another typical trajectory, but 'a great many' went into the Church.[49] Some of Richardson's ancestors, like his grandfather John Richardson, were Trinity men, as were many of his own contemporaries. His near neighbour, Stewart Blacker, became Dean of Leighlin. Charles Stewart, an early eighteenth-century predecessor in the College living of Clonfeacle, anticipated Richardson's academic trajectory of BA, MA, BD and DD, but had risen to be Provost.[50] Thomas Torrens, brother of his old headmaster, followed a similar academic and clerical trajectory to Richardson.[51] Thomas Campbell entered Trinity earlier than Richardson (in 1752) but their careers overlapped, as did their subsequent clerical and writing activities. Campbell became a Tyrone rector and wrote a 'philosophical' survey of southern Ireland in 1778.[52] But proximity could also bring problems.

Walter Ricky had graduated with a BA in 1778 and was curate at Clonfeacle when Richardson arrived in the parish in 1783. Relations were initially congenial but, as we shall see, sparks flew when Ricky crossed Richardson.[53] Ricky was at still Trinity during Richardson's fellowship and, if taught by him, this could explain the virulence of their quarrel. We do know two men who were definitely taught by Richardson and he exploited the connection to the hilt afterwards. Richardson's *élèves* included the Reverend William Bruce, a Presbyterian exception to the Anglican monopoly of Trinity, who later became Richardson's main contact in Belfast. Isaac Corry, first post-union Irish Chancellor, was also an ex-pupil. Alongside their official careers, both men were socially well connected and interested in science and improvement, and Richardson dedicated pamphlets to each.[54]

The Trinity education was described as encouraging 'the vigorous shoots of genius' for men destined to be leaders in their fields.[55] Some of Richardson's contemporaries, like William Magee, achieved high positions in the Church; others, like Arthur Wolfe, attained distinction in the judiciary; and still others, like Parsons, gained reputations in science and agricultural improvement. Though Richardson failed to progress in the Church, his dominant passions for geology and agricultural improvement were well catered for. A Trinity graduate, Dr Samuel Madden, had helped establish the Dublin Society in 1731, 'for the improvement of husbandry, manufactures and other useful arts', which offered money incentives for improvement schemes. 'Premium Madden' later inaugurated a similar practice for students who excelled in their examinations.[56] Though geology was not formally taught in Richardson's time, there was a great interest in the subject. A museum was established in 1777 to house various artefacts collected during Captain Cook's Pacific voyage and also geological and zoological specimens.[57]

Richardson left Trinity in 1783 to take a College living at Clonfeacle. It was customary for vacant livings to be offered to clerical fellows in order of seniority, leaving them free to marry. As Richardson did not marry until 1785, a more likely explanation was the financial and social inducement. Clonfeacle was a large parish on the Armagh–Tyrone border worth £1,123 annually, with over 500 acres of glebe, making it relatively lucrative.[58] Though not in the top bracket – at least six clerical livings were worth £1,500 a year – it was prestigious: only seven or eight livings brought over £1,000, more than double what a senior fellow could ever hope for.[59]

We have Richardson's own testimony that he continued growing physically until he was twenty-three. This would have been in 1763, the year he gained his BA; however, his intellectual growth did not stop there. The combination of a childhood of maternal indulgence followed by years in a male educational institution which encouraged confrontational *viva voce* declamation produced a robust individual with the courage of his own convictions. This atmosphere would have been congenial. Destined for the Church, Richardson's thinking was shaped in an environment where natural theology complemented classical learning. Trinity toughened up a rather delicate child. This self-belief gave Richardson confidence to express his opinions, but could also produce intolerance for anyone with the temerity to disagree with him. Fellowship signified a social position but, perhaps because he could not take this reputation for granted, he guarded it manically and saw slights where none existed. These traits came out strongly in his writings, but he raised his voice to his family and parishioners long before he lifted his pen.

Family tensions

Insights into this can be gained from the 1780s correspondence of Richardson's Coleraine cousin, Elizabeth Heyland, with her son Rowley. Richardson was well aware of his considerable social influence, but had a very difficult and cantankerous personality, ready to take umbrage, particularly over money. Rowley Heyland was apprenticed to another relative, the Dublin solicitor Dominick McCausland. Unsupported by her 'crabbit' husband, Richard, and financially insecure, Elizabeth needed Richardson's help, advice and his Dublin connections. But she worried constantly about his touchy sense of propriety. Richardson undertook to support young Rowley, but money was not forthcoming when needed. Elizabeth cautioned her son not to press too hard for money, 'supposing Dr Richardson did not think you would want money so soon'.[60] The impecunious Rowley was understandably nervous at meeting his imperious uncle. His mother had initially told him to write but, fearing Richardson might construe a written approach as implying he could not be taken at his word, then advised Rowley to meet him personally. Rowley seems to have baulked at this, as his overwrought mother said that she was 'greatly vexed and surprised at your not seeing Dr Richardson'. She then assured him that 'I don't believe he [Richardson] ever heard anything to your disadvantage', and explained that she would not have suggested the letter except 'that Mrs McCausland insisted on it'. In a pattern that would frequently recur in his scientific spats, the more Richardson's prickly pride was activated, the more obdurate he became. In despair Elizabeth told Rowley, 'if you don't see him and their [*sic*] is nothing done don't write to him again but draw a bill upon your father payable to your master at the Assizes'.[61] As Richardson was settled in Clonfeacle and had doubled his income, it seems that this situation arose less from his own financial circumstances than embarrassment at being importuned by straitened relatives.

Richardson's own outgoings increased on 9 May 1785 when he married Hannah, the daughter of Mark and 'Black Bess' McCausland. As Richardson's new mother-in-law was a Heyland, this maintained the McCausland–Heyland–Richardson connection, as his own mother was Sally Heyland before she married Charles Richardson.[62] Hannah was about ten years younger than William, whom she outlived by almost twenty years.[63] Elizabeth Heyland was 'greatly pleased with Dr Richardson's match' and the wedding was held in Rush Hall, a McCausland property near Newtownlimavady in county Londonderry, from where the party went to Portrush.[64] Now a popular seaside resort, Portrush then, with its bracing Atlantic air, pristine coastline and sublime natural formations, was a summer retreat for the genteel. It was a natural place for the first weeks of married life. Richardson had already leased property there from the Earl of Antrim and, after his marriage, spent every summer in Portrush. The town was close to his relatives in Coleraine, where he had inherited more property, from the Reverend Robert Heyland.[65] The newly-weds were greeted by 'constant crowds of people'. Elizabeth urged Rowley to come to Portrush, as 'never did you see a happier set of people'.[66] The joy, however, was mingled with sadness and ominous signs for the future.

One day in June 1785 the party went boating, crossing the mouth of Lough Foyle into Greencastle in county Donegal, when an Atlantic storm blew up. Richardson's friend, the Reverend Gardner Young's pregnant wife, became so seasick that she suffered a miscarriage. Affected by this, Richardson ruefully remarked that his own wife 'shall boat no more'. By the autumn Hannah too was pregnant, but became so ill that Elizabeth Heyland warned Rowley that Richardson must not be approached 'on any account'. The fact that Elizabeth kept her distance speaks volumes about Richardson. However, by January, as Hannah remained 'very ill', he took the initiative and asked Elizabeth to help.[67] The outcome is not recorded but, as a child of the Richardsons, Charles, died young, this must have been the unhappy conclusion to a difficult time.[68] If so, Hannah's request that Rowley buy her 'a small soft brush for the child's head' assumes additional poignancy.[69] Family records suggest that William and Hannah also had a girl, Bessie, who 'died unmarried', although there is no other mention of her. Four (male) children survived into adulthood. The eldest was Marcus (a common McCausland Christian name), who became a captain in the 68th Regiment. The second son, Arthur Wolfe Richardson, was born in 1794 and named for his father's close friend and political collaborator Lord Kilwarden. Arthur followed his father into Trinity and the Church, becoming Perpetual Curate of Moy (part of Clonfeacle), but surviving his father by only two years. In 1797, a third son, William, was born and, as we shall see, inherited as well as his name the worst aspects of his father's personality. Richardson's youngest son, another Charles, also went into the Church, succeeding his brother Arthur at Moy, and married into another local family, the Kings of Newtownlimavady.[70] In his dealing with relatives, Richardson's self-righteousness, combative obstinacy and intolerance of contradiction emerged, traits which also coloured relations with parishioners and even his own children.

Rector of Clonfeacle

Clonfeacle had ancient ecclesiastical and military significance. The name is an Anglicisation of the Irish *Cluain Fiaclul*, meaning 'the church of the tooth', because a dental relic of Saint Patrick was supposedly preserved there.[71] Its location alongside the River Blackwater, which forms the Armagh–Tyrone border, gave it strategic importance. During Elizabeth I's reign Hugh O'Neill had defeated English forces nearby at the Battle of the Yellow Ford. The area was also disturbed when native Irish forces achieved victory over a Scots parliamentarian army in 1646 at Benburb, where Richardson's church stood.[72] Richardson certainly fought tooth and nail with parishioners and with his curate and fellow Trinity graduate Walter Ricky.

On arriving in the parish he wanted to make the place suit his tastes. The church was built between 1618 and 1622 in the style known as 'Planter's Gothic', and the glebe house was erected in 1751 at a cost of £1,189.[73] In 1784 Richardson set about improving his glebe house, memorialising Primate Robinson, no mean improver himself, to approve the cost of unspecified alterations. A similar memorial was sent in 1802 to the then Primate, Archbishop William Stuart, to harness his scheme for interest-free loans for clergymen to build glebe houses.[74] This latter memorial reflected Richardson's ambitions, noting that inadequate existing buildings caused 'great inconvenience from want of sufficient accommodation for servants'. He planned to raise the walls of his coach house, saddle room and potato house, creating comfortable servants' quarters, and to build a servants' 'necessary house'. Always attentive to his own comforts, he demolished walls which impeded the 'benefit of air and light', and built new walls to protect his garden and a lime kiln. He followed this memorial for £220 with a revision adding a further £50 for replacing shingles on the barn roof with slates, to which Stuart consented.[75] As befitted his status, Richardson's servants included a butler, a land steward and a coachman.[76]

These improvements made Clonfeacle a haven for relatives to visit, thus increasing his position in the family. It was a comfortable winter equivalent to Portrush, with the added frisson of excitement from Richardson's elite social connections. In December 1784, Elizabeth Heyland had 'never spent time more agreeably', as the visit included a dinner at Caledon House, where she hobnobbed with Richardson's neighbour, Lord Caledon. This impressed the impecunious Elizabeth, whose only disappointment was that Caledon retired after dinner, a martyr to gout. The Clonfeacle effect extended to her 'crabbit' husband, who was in better spirits after a visit than for many months. A further invitation to wealthy Heyland relatives at Glenoak near Crumlin, county Antrim, was delayed due to a prodigious snowfall, but Elizabeth's party were happy to stay at Clonfeacle. At Christmas and New Year 1786–87 she 'spent her time very pleasantly' and 'met with the greatest friendship and affection from this family but if possible I meet with more of it every day'.[77]

Though prepared to lavish church money on his house, Richardson was parsimonious with his parishioners, especially his curate, Walter Ricky. He had an acrimonious row with Ricky, who in 1807 published a pamphlet provocatively headed with a text from Proverbs: 'devise not evil against thy neighbour'. In it he complained bitterly of Richardson's 'extraordinary conduct' towards him.[78] This clerical dispute originated less in the service of God than of Mammon: Richardson had refused to raise Ricky's salary. This was a more general problem leading to the 'curate's revolt'. The British parliament raised curates' salaries but Irish curates complained that clergymen would dismiss them rather than increase their salary. Irish legislation of 1800 was designed to circumvent this. It gave bishops the right to license curates who had not been nominated by the incumbent and to set pay rates, with archbishops having the final say.[79] As Ricky's curacy predated Richardson's arrival, he was not the rector's nominee. This made the relationship less unequal, though unreflective of the gap which existed at Trinity between Richardson and Ricky. Despite originating in a general problem, the row soon became very personal. Ricky wrote: 'I had offended him [Richardson] most grievously for I had presumed to look for what the law allowed me and what his Grace (the Primate) thought I was entitled to'. Ricky alleged that Richardson accused him of writing scandalous anonymous letters about his wife, in 'the grossest language' and charging her with 'the foulest of crimes', as a device to get him dismissed.[80]

Ricky presented Richardson as a Machiavellian manipulator. He claimed that he invited gentlemen for dinner – 'the doctor's usual mode of bribing those he wished to swallow his calumny' – and instituted an investigative committee 'whose ears he purchased with beef and port'. Sure-footed about legal procedures, Richardson blackened Ricky's name but adroitly avoided legal action, hoping that 'this pestilent curate, who dared to apply to the Primate for the increased salary will quit his cure'. In the event, Richardson was forced to pay the increased salary but Ricky unwisely wrote a letter demanding an apology. Richardson peremptorily refused even to open this letter, telling the Primate that he had made up his mind to assume the parish duties himself. However, the rector soon 'seemed to tire of well doing' and had someone waiting 'at his beck' to 'slide into the Cure'. Richardson subsequently refused to speak to Ricky, who, finding this unbearable, confronted him after Sunday service suggesting that they should forget 'what is past … and live… as clergymen should'. But this olive branch, 'even on the threshold of the house of God', received the abrupt Richardsonian reply 'I never will!!!' Rickey ended with the remarkable claim that Richardson fabricated the letters himself, but the rancorous rector had the last word, appointing a new curate in 1807. Richardson later proved that the offending letters were written by the butler of the rector of Killyman, perhaps unsurprisingly, as Richardson's combative personality meant that his relations with neighbouring clergymen were 'notorious'.[81]

Tensions with parishioners simmered at vestry meetings. Clonfeacle was a straggling parish that included part of Armagh as well as of Tyrone, but the church was at Benburb, at the county border, which apparently made attendance difficult for many Tyrone parishioners. Money to build a chapel of ease at Moy in Tyrone had been obtained from the Board of First Fruits. This, the parishioners emphasised, was to be 'a *free* chapel, no private pews, but open to all comers'. But the grant was judged insufficient and the Tyrone vestrymen applied to levy 1*d* per acre on the entire parish to supplement the grant. Richardson was very reluctant to agree to this, fearing, it was said, having to pay another curate. At one such vestry meeting he tried to head the vestrymen off, resorting to legal protocol, claiming that the meeting was illegal without prior notice. However, the parishioners called his bluff by asking for a week's adjournment in case Richardson wanted to get a legal opinion. In a typical reaction to contradiction, he refused the adjournment, complained about 'many strange faces in the Church and none of the County of Armagh', snapped the vestry book shut in their faces and stormed out of the building.[82] Again he had the last word: when the Moy chapel was eventually built his second son, Arthur, became perpetual curate in 1819.[83] In fairness, the problems were not all Richardson's making: a successor, William Davenport, blew his brains out in 1822 after a year at Clonfeacle.[84] However, Richardson's rebarbative personality meant conflict in any situation, public or private.

Fatherhood

His third son, William, seems to have inherited his father's volatile combination of pugnacity and sensitivity and his predilection for attracting attention. Richardson utterly indulged his namesake and could see no wrong in him despite much evidence to the contrary. Young William was launched into a military career as an officer cadet in the Royal Military Academy at Woolwich and was provided with a private tutor to ensure he passed his examinations. But a 'series of misbehaviours' convinced his superiors that his 'natural disposition [was] quite unsuited to his situation here'. The colonel-in-charge explained to his father how 'Mr. Richardson's behaviour has been so truly vexatious, imprudent and insolent' that he must be removed to avoid the disgrace of expulsion. One of his more provocative japes was to appear on parade with sideburns and moustache pencilled on his face. Disappointed but not deterred, Richardson pressed on with his determination for his son to have a military career and purchased a commission in the 38th Regiment. But his insubordinate son ran into trouble again. He narrowly avoided a duel with an officer who had 'calumniated my caracter [*sic*] in such a manner as I was obliged to call him out'. A further parental purchase brought William into the 83rd Regiment, where he matured and got on better with his fellows. But he was posted to Ceylon, where he married a widow in 1822 but died the following year.[85] By contrast with the indulgence of the youngest, the eldest son, Marcus, felt the force of parental fury in 1811 when he announced his engagement to Mary Ann Gage.

The Gages were an important landed family, owning Rathlin Island and other property, and, as such, Marcus had apparently made a good match. However, he had made the announcement without having obtained prior parental approval. Richardson was mortified by this filial rebellion, and also dreaded the financial implications, less for the young couple's start in life than for what this would signify about him. He refused his consent 'in the most decided manner', telling Mary Ann's mother, Barbara, one of the Coleraine Richardsons and thus a relative, that 'from the improvidence of my son, his captain's pay is not adequate to his own support'. Richardson's friend, the Reverend Gardner Young, mediated between the engaged couple and their respective families. Young relayed Mary Ann's concerns about such 'decided opposition' emanating from Clonfeacle, especially when she and Marcus had documents proving that consent had in fact been obtained. Mary Ann had actually said that 'great duplicity towards her proceeds from the house of Clonfeacle'. Knowing his man, Young had toned that down. He tried a conciliatory letter to Richardson, but received an explosive riposte which demolished every argument. Young complained, accurately, that 'the Doctor has a passion for indulging his pen which, though it may be life to him, is death to me', but kept copies of the letters in case the dispute went to law.[86]

The correspondence does not survive but it probably included letters from a Mrs Boyle, whom Marcus represented as his father's friend. Richardson denied this, claiming that his son had 'seduced' Boyle by misrepresenting Mary Ann's dowry as £2,000 and 'persuaded her to use her influence with me'. Letters from Richardson's wife were also involved. Richardson's reply to Young mentions letters from Hannah about the projected union, which she had written '*without my concurrence or knowledge*'. Furious, Richardson responded with an unanswerable line of forensic questioning. How could Marcus assume that Boyle's letters conveyed consent then subsequently ask for his father's agreement? If the couple persisted in their engagement plan, they did so 'under the pretence of a contract obviously patched up as an excuse for predetermined disobedience'. It was, Richardson fumed, without any sense of irony at his indulgence of William, 'highly aggravating' and an abuse of the independent situation in which he had placed Marcus, meaning the purchase of a commission. Reading between the lines, the young couple, like Gardner Young and Elizabeth Heyland, had correctly anticipated Richardson's overreaction. At bottom this response was probably motivated by social embarrassment that a Trinity fellow's son could not provide for the daughter of one of his Coleraine relatives. Richardson may also have felt that Marcus did not cut the figure he would have expected from his eldest son. Walter Ricky claimed that Marcus had been so bullied at Dungannon Royal School that his father withdrew him. Fortunately for Marcus and his fiancée, Mrs Gage pragmatically cut through the convoluted attempts to navigate around Richardson. She recognised the financial problem, but understood that events had passed the point of no return, gave her consent and the couple were married on 18 May 1811.[87]

Richardson's character

Richardson's early life and family relationships therefore reveal a highly intelligent and capable man, but with a difficult personality to be tiptoed around. He was pugnacious and tenacious in equal measure. Riven by contradictions, throughout life he remained utterly convinced of his own rectitude; yet, at the same time, he could be completely blind to the viewpoint of another. A superior attitude masked his inner social insecurity. He felt a constant need to prove himself and had the dialectical skills to do it. Trinity disputations gave him the ability to construct a convincing argument and the inductive reasoning that he taught his pupils enabled him to see beyond the initial appearance of a problem and understand its possible consequences in the abstract. These character traits, revealed in his private and family life, appear in abundance in his public involvements. However, Richardson's actions should not obscure the fact that he was, by any estimation, an extraordinary man living in extraordinary times, as his political conduct shows.

Notes

1 'Virtus Paret Robier', the Richardson family of county Tyrone and Somerset, Coleraine, PRONI, Maxwell Given papers, D2096/1/19A.
2 J. Dubordieu (ed.), *A statistical survey of the county of Antrim* (Dublin, 1812), vol. 2, p. 101.
3 The Richardson family of county Tyrone and Somerset, Coleraine, PRONI, Maxwell Given papers, D2096/1/19A.
4 PRONI, Maxwell Given papers, D2096/1/19A; A. I. Young, *Three hundred years in Innishowen* (Belfast, 1919), p. 112; W. R. Young, *The fighters of Derry* (London, 1932), pp. 187, 223.
5 E. M. Johnston-Liik (ed.), *History of the Irish parliament* (Belfast, 2002), vol. 6, pp. 155, 157–8.
6 Notes on the Richardson family, PRONI, Maxwell Given papers, D2096/1/19B; Young, *Fighters of Derry*, pp. 112, 223; W. E. C. Fleming (ed.), *Armagh clergy, 1800–2000* (Dundalk, 2000), pp. 298–9; I am grateful to Jennifer Cunningham of the Coleraine Historical Society for information on the Richardson family.
7 Johnston-Liik, *Irish parliament*, vol. 6, p. 158.
8 S. J. Connolly, *Religion, law and power: the making of Protestant Ireland, 1660–1760* (Oxford, 1992), pp. 299–300; Fleming, *Armagh clergy*, p. 298.

9 Richardson to Greenough, 10 October 1810, UCL, Additional Greenough papers, 1449.
10 Young, *Innishowen*, p. 112.
11 PRONI, Heyland papers, D4085/E/5; Notes on the Richardson family, PRONI, Maxwell Given papers, D2096/1/19B.
12 Richardson to Greenough, 29 November 1812, UCL, Additional Greenough papers, 1475.
13 Young, *Innishowen*, pp. 209–11; Young, *Fighters of Derry*, 186; J. V. Luce, *Trinity College Dublin: the first 400 years* (Dublin, 1992), p. 51.
14 T. Barnard, *A new anatomy of Ireland: the Irish Protestants, 1649–1770* (New Haven and London, 2003), pp. 106–11.
15 C. Maxwell, *A history of Trinity College Dublin, 1591–1892* (Dublin, 1946), p. 132.
16 M. Elliott, *Wolfe Tone: prophet of Irish independence* (New Haven and London, 1989), p. 10.
17 Maxwell, *Trinity College*, pp. 131–2, 142; Elliott, *Wolfe Tone*, pp. 14–15; J. Hill, 'Corporatist ideology and practice in Ireland, 1660–1800', in S. J. Connolly (ed.), *Political ideas in eighteenth-century Ireland* (Dublin, 2000), p. 67.
18 Luce, *The first 400 years*, p. 7.
19 Maxwell, *Trinity College*, p. 148.
20 Luce, *The first 400 years*, p. 62.
21 R. B. McDowell and D. A. Webb, *Trinity College Dublin, 1592–1952: an academic history* (Cambridge, 1982), p. 70.
22 Luce, *The first 400 years*, p. 62.
23 Sir Jonah Barrington, *Personal sketches* (London, 1830), p. 39; Johnson-Liik, *Irish parliament*, vol. 3, p. 136.
24 Elliott, *Wolfe Tone*, pp. 31–2; Luce, *The first 400 years*, p. 69 note 1.
25 Barrington, *Personal sketches*, p. 39.
26 Luce, *The first 400 years*, pp. 50–1.
27 J. W. Foster, 'Encountering traditions', in J. W. Foster (ed.), *Nature in Ireland: a scientific and cultural history* (Dublin, 1997), p. 51.
28 C. W. P. Mac Arthur, 'Richardson, William (1740–1820)', *Oxford dictionary of national biography* (Oxford, 2004), online edition, January 2008, www.oxforddnb.com/view/article/23591, accessed 26 August 2011.
29 Maxwell, *Trinity College*, pp. 127, 141; P. J. Jupp, 'Dr. Duigenan reconsidered', in S. Wichert (ed.), *From the United Irishmen to twentieth-century unionism* (Dublin, 2004), pp. 80–1.
30 Maxwell, *Trinity College*, pp. 120, 126, 130–1; Luce, *The first 400 years*, p. 51; McDowell and Webb, *Trinity College Dublin*, p. 52.
31 See, for example, W. Richardson, *Memoir on fiorin grass to the Belfast Literary Society 1 March 1808* (Belfast, 1808), frontispiece.
32 Luce, *The first 400 years*, p. 61.
33 G. D. Burtchaell and T. U. Sadlier (eds), *Alumni Dublinenses, 1593–1860* (Dublin, 1935), vols 8–9, p. 703; Maxwell, *Trinity College*, pp. 115–20, 201; McDowell and Webb, *Trinity College Dublin*, p. 52.
34 Cited in Maxwell, *Trinity College*, pp. 115–20; see also McDowell and Webb, *Trinity College Dublin*, p. 52.
35 McDowell and Webb, *Trinity College Dublin*, pp. 64, 66, 70; Maxwell, *Trinity College*, pp. 122–3; 148–9.
36 Barrington, *Personal sketches*, p. 140.

37 Richardson to Thomas Knox, 12 July 1791, PRONI, Abercorn papers D623/A/133/25.

38 Connolly, *Religion, law and power*, pp. 239–40; Elliott, *Wolfe Tone*, p. 20.

39 T. Bartlett (ed.), *The life of Theobald Wolfe Tone* (Dublin, 1998), p. 18.

40 Richardson to Greenough, 27 January 1808, UCL, Additional Greenough papers, 1417; Elliott, *Wolfe Tone*, pp. 16–17.

41 D. Budd and R. Hinds, *The 'Hist' and Edmund Burke's club* (Dublin, 1997), pp. 13, 17, 39.

42 Richardson to Greenough, 27 December 1807, UCL, Additional Greenough papers, 1405.

43 Barrington, *Personal sketches*, p. 210.

44 Ronan Keane, 'Toler, John, first Earl of Norbury (1745–1831)', *Oxford dictionary of national biography* (Oxford, 2004), online edition, January 2008, www.oxforddnb.com/view/article/27498, accessed 9 December 2011.

45 W. N. Osborough, 'Wolfe, Arthur, first Viscount Kilwarden (1739–1803)', *Oxford dictionary of national biography* (Oxford, 2004), online edition, May 2009, www.oxforddnb.com/view/article/29830, accessed 9 December 2011.

46 Burtchaell and Sadlier, *Alumni Dublinenses*, pp. 287–8, 655, 815; Ann C. Kavenaugh, *John Fitzgibbon, Earl of Clare* (Dublin, 1997), p. 14.

47 Burtchaell and Sadlier, *Alumni Dublinenses*, p. 258.

48 Johnson-Liik, *Irish parliament*, vol. 4, p. 26; W. Richardson, *An essay on the improvement of the great flow bogs of Ireland particularly the Bog of Allen and the Montiaghs of the North … in a letter addressed to the Grand Juries of Antrim, Armagh and Tyrone* (Dublin, 1807), frontispiece.

49 Cited in Maxwell, *Trinity College*, p. 153.

50 Burtchaell and Sadlier, *Alumni Dublinenses*, pp. 70, 790.

51 Maxwell, *Trinity College*, pp. 130–1; The Richardson family of county Tyrone and Somerset, Coleraine, also the Torrens family of Somerset, Coleraine, PRONI, Maxwell Given papers, D2096/1/19A.

52 Burtchaell and Sadlier, *Alumni Dublinenses*, p. 130; Maxwell, *Trinity College*, p. 131.

53 Burtchaell and Sadlier, *Alumni Dublinenses*, p. 704; W. Ricky, *The extraordinary conduct of the Rev. Dr. Richardson, rector of the parish of Clonfeacle … towards the curate of the same parish* (Dublin, 1807), pp. 5, 13, 16–17, 27.

54 Richardson, *Memoir on fiorin grass*, preface; W. Richardson, *Letter to the Rt Hon. Isaac Corry, containing an epitome of some of the most curious and important properties of Irish fiorin, or fyoreen grass, with proofs that the facts by which they have been established, are fairly stated, and that the author is not mad* (Belfast, 1809); Johnston-Liik, *Irish parliament*, vol. 3, pp. 513–15.

55 Elliott, *Wolfe Tone*, p. 41.

56 Maxwell, *Trinity College*, pp. 154–6; Luce, *The first 400 years*, p. 63 note 10.

57 McDowell and Webb, *Trinity College Dublin*, p. 195; P. N. Wyse Jackson, 'Tumultuous times: geology in Ireland, 1740–1816', in P. N. Wyse Jackson (ed.), *Science and engineering in Ireland in 1798* (Dublin, 2000), p. 38.

58 Richardson to Greenough, 27 January 1808, UCL, Additional Greenough papers, 1417; Mac Arthur, 'Richardson, William (1740–1820)', p. 863; 'State of the glebe house of Clonfeacle', PRONI, Armagh diocesan papers, DIO/32/C/5/1.

59 Luce, *The first 400 years*, pp. 86–7; McDowell and Webb, *Trinity College Dublin*, p. 102; PRONI, Armagh diocesan papers, DIO/32/C/5/1.

60 Elizabeth Heyland to Rowley Heyland, 13 July 1784, PRONI, Heyland papers, D4058/D/1/5, D4085/E/5.
61 Elizabeth Heyland to Rowley Heyland, 28 December 1784, 10 January 1785, PRONI, Heyland papers, D4085/D/1/10, 11.
62 Fleming, *Armagh clergy*, pp. 298–9; Genealogy of the Richardson family, PRONI, Maxwell Given papers, D2096/1/19A.
63 Mac Arthur, 'Richardson'; Prerogative grant book, 1839, folio 174A, NAI (I am grateful to Dr Brian Trainor of the Ulster Historical Foundation for this information).
64 Elizabeth Heyland to Rowley Heyland, 12 April, 4 May, 30 May 1785, PRONI, Heyland papers, D4085/D/61/16, 17, 18.
65 Counterpart lease between the Earl of Antrim and William Richardson, 1 November 1782, PRONI, Earl of Antrim papers, D2977/3A/2/35/14; Last will and codicils of the Rev. Robert Heyland, 30 June 1796, PRONI, Hezlett papers, D668/49/2; Elizabeth Heyland to Rowley Heyland, 13 July 1784, PRONI, Heyland papers, D4058/D/1/5.
66 Elizabeth Heyland to Rowley Heyland, 30 May, 26 June 1785, PRONI, Heyland papers, D4085/1/18, 19.
67 Elizabeth Heyland to Rowley Heyland, 26 June, 20 November 1785, 9 January 1786, PRONI, Heyland papers, D4085/D/1/19, 22, 23.
68 Fleming, *Armagh clergy*, pp. 298–9.
69 Elizabeth Heyland to Rowley Heyland, 13 December 1786, PRONI, Heyland papers, D4085/D/1/28.
70 George W. Phipps to General Trotter, 3 February 1813; Colonel Mudge to Richardson, 8 February 1813; [Young] William Richardson, Winchester, to Greenough, 30 June 1814, UCL, Additional Greenough papers, 1478, 1480, 1486; Fleming, *Armagh clergy*, pp. 298–9.
71 A. Day and P. McWilliams (eds), *Ordnance Survey memoirs of Ireland, vol. 1: parishes of county Armagh, 1835–8* (Belfast, 1990), p. 23; S. Lewis, *Counties Fermanagh and Tyrone: a topographical dictionary* (London, 1837; new edition Belfast, 2004), pp. 55–6.
72 H. Simms, 'Violence in county Armagh', in B. Mac Cuarta (ed.), *Ulster 1641: aspects of the rising* (Belfast, 1997), p. 132; M. O'Riordan, 'The native Ulster mentalité as revealed in Gaelic sources, 1600–1650', in Mac Cuarta (ed.), *Ulster 1641*, p. 87.
73 P. Larmour and S. McBride, 'Buildings and faith: church building from medieval to modern', in R. Gillespie and W. G. Neely (eds), *The laity and the Church of Ireland, 1000–2000* (Dublin, 2002), p. 307; Fleming, *Armagh clergy*, p. 306.
74 A. P. W. Malcomson, *Archbishop Charles Agar: churchmanship and politics in Ireland, 1760–1810* (Dublin, 2002), pp. 258–61.
75 Draft memorials, 28 January 1802, 1 July 1803, PRONI, Armagh diocesan papers, DIO4/32/C/10/5/3; Ordnance Survey map 1838, PRONI, OS1/6/61.
76 Richardson, *Letter to the Rt. Hon. Isaac Corry*, pp. 21, 24, 29.
77 Elizabeth Heyland to Rowley Heyland, 6, 13 December 1784, 9 January 1785, PRONI, Heyland papers, D4085/D/1/19, 28, 30.
78 Ricky, *Extraordinary conduct*.
79 Malcomson, *Archbishop Charles Agar*, pp. 578–81
80 Ricky, *Extraordinary conduct*, pp. 5, 13, 16–17, 27.
81 Ricky, *Extraordinary conduct*, pp. 18–27, 29; W. Ricky, *Dr Richardson's calumny refuted* (Dungannon, 1808), pp. 4, 10; Fleming, *Armagh clergy*, pp. 298–9.

82 Memorial to the Primate, n.d. [watermarked 1811], PRONI, Armagh diocesan papers, DI04/32/C/10/7/1.

83 Fleming, *Armagh clergy*, p. 298.

84 McDowell and Webb, *Trinity College Dublin*, p. 526 note 19.

85 George W. Phipps to General Trotter, 3 February 1813, Colonel Mudge to Richardson, 8 February 1813, [Young] William Richardson, Winchester, to Greenough, 30 June 1814, UCL, Additional Greenough papers, 1478, 1480, 1486.

86 Richardson to Barbara Gage, 20 March 1811, Young to Richardson, 13 April 1811, Young to Robert Gage, 13 April 1811, PRONI, Gage papers, T1883/1, 2, 4.

87 Richardson to Young, 16 April 1811, Young to Richardson, 13 April 1811, PRONI, Gage papers, T1883/3; Ricky, *Dr Richardson's calumny refuted*, p. 9.

2

William Richardson: popular loyalism and the politics of Protestant Ascendancy

Clonfeacle straddles the River Blackwater and its rector, like Janus, the Roman god of bridges, looked towards the future as well as back to the past. But, unlike the stony classical deity, William Richardson could change with the times. Compared to loyalism in Britain, the study of its Irish equivalent of the 1790s has languished in the historiographical shadow of radicalism and nationalism, even though the eighteenth century is seen as the 'Protestant century'.[1] The neglect of this pivotal decade partly reflects a malaise which once affected British history: that those who sought political change are more interesting and worthy subjects of study than conservatives who resisted it. But there is also a conceptual problem. This concerns the tendency of Irish historians to use a 'history from below' approach to understand radicalism but to apply a top-down analysis to its opponents.[2] One result is a conspiratorial interpretation of loyalism which sees it as deliberately fostered by the Irish government and Ascendancy grandees to introduce sectarianism as a counter-revolutionary strategy to the United Irishmen.[3] This obscures the dynamic relationship between popular loyalism and elite Protestant Ascendancy politics.

Interpretations which depict Irish loyalism as the dull, even dark, backdrop against which enlightened radicalism shines unfortunately assume that plebeian loyalists were unthinking automatons, hidebound with an instinctive anti-Catholicism and blindly deferential to their social superiors, like an Hibernian version of E. P. Thompson's (subsequently retracted) depiction of English loyalists as flag-saluting, foreigner-hating ciphers of the elite.[4] Recent work on Irish loyalism has begun to reveal that, like its British equivalent, it was more nuanced, more socially and politically variegated and far less deferential than previously thought.[5] For the Protestant elite, there are good works on eighteenth-century Ireland and important biographical studies of political, landed and ecclesiastical political figures, like John Foster and Archbishop Agar. Corporate Ireland, another bastion of Ascendancy, has also received timely attention, as have publicists like Edward Newenham and Sir Richard Musgrave.[6] However, further work is required, particularly concerning the dynamics between the Ascendancy elite and plebeian loyalists. This crucial relationship is often obscured by a dearth of sources concerning those who could move between both worlds. This is not so, however, for Richardson, whose lengthy private correspondence with George Bellas Greenough reveals the former's intermediary role. Richardson met Greenough in July 1806, when he and Humphry Davy visited the Giant's Causeway during a geological tour. Greenough was a scientist and President of London's Geological Society but was also MP for Gatton. His politics instinctively favoured Whiggish reform and his journal reveals a sensibility offended by the poverty and political powerlessness of the Catholic Irish.[7] Though Greenough's side of the correspondence does not survive, Richardson's letters reveal extensive exchanges on topics related to loyalism, Orangeism and the difficulties of putting the Ascendancy case in Britain. This chapter is organised around these interlinked themes. It examines Richardson's role as intermediary between elite and plebeian in the promotion of loyalism and asks why his support for Catholic relief gave way to an uncompromising 'ultra' position. Richardson's manipulation of information illuminates the real dilemmas faced by Protestants after union and how the loyalism of the 1790s could be a liability in the 1800s. Amateur militarism was a key component.

Yeomanry and militia

Militarism and loyalism were joined at the hip in a long-established 'Protestant defence tradition'[8] which Richardson's own ancestors at the siege of Derry would have readily understood. This produced defence forces like Williamite country associations, militias official and voluntary, the Irish Volunteers of 1782 and the yeomanry of 1796, temporary forces embodied from civilians in wartime or when insurrection threatened. These amateur military forces involved different levels of Protestant society, typically being raised and commanded by the landed elite and manned by their tenants. A politics of participation was operative, but it fluctuated between willing involvement and obstinate independence, according to the perceived threat. Ideology played a role: did loyalty signify automatic readiness to bear arms against the King's enemies or did raising military bodies from civilians infringe the principle of consent enshrined in Whig thinking? Such tensions defined the Volunteers of the 1780s, when, in default of parliamentary funds for a militia, men banded together to defend their localities against invasion but refused to put themselves in government control or pay.[9] In 1793 a new Irish militia was raised. Unlike any similar eighteenth-century force, it included Catholics in the ranks. It also had an element of compulsion, being initially raised by ballot, with each parish liable for a manpower quota, but wealthy people could pay a substitute or buy their way out till the next ballot. The ballot caused protests, partly for ideological reasons but also because balloted men were paid for by parish cess and became a charge on landholders. Balloting was designed for a part-time force but the militia necessarily became full time and other recruitment methods were tried as the full establishment was not reached.[10] In 1795, with the state of the country worsening, legislation to augment the militia reopened the controversy and involved Richardson.

Militia colonels were landed magnates and their county regiments a matter of pride as well as electoral patronage (they could nominate officers). In Tyrone, the Marquis of Abercorn had successfully lobbied for his regiment to bear the prefix 'Royal'.[11] But tensions arose when central regulations cut across landed pride, and the ramifications were felt severely in Clonfeacle, where protests exposed the delicate social processes which would soon be necessary in organising loyalism in yeomanry corps. Richardson was embroiled in a dispute between his parishioners and the gentry and found himself in the uncomfortable situation of standing up to Abercorn, one of Ireland's most powerful aristocrats. His response anticipates later rapprochements between elite and plebeian.

To understand the problem, it is necessary to sketch in the context. Richardson's own property lay in Tyrone and he supported Abercorn's political interest. Nicknamed 'Magnifico', Abercorn had estates in England, Scotland as well as Ireland, where he owned 26,500 acres in Tyrone and 10,000 in Donegal. A personal friend of Pitt, he ambitiously sought the Irish viceroyalty, conceiving himself as a new political power independent of the ultra-Protestant clique which supported the government. Tyrone was a large, open constituency, with an electorate of around 3,500, which included many independent-minded Presbyterians, and had increased to 20,000 after the enfranchisement of forty-shilling freeholders in 1793. Tyrone's representation was divided between James Stewart, returned on the Presbyterian interest, and Abercorn's supporter and Richardson's friend Thomas Knox of Dungannon. He had managed Abercorn's Tyrone interest and Richardson's Trinity connections were exploited to the same ends.[12] This mutually beneficial scenario disintegrated in 1794 when Abercorn and Knox quarrelled over the former's resignation as lieutenant-colonel of the Tyrone militia, and his command devolved to Nathaniel Montgomery-Moore. This bitter row strained, but did not break, the fabric of Abercorn's interest. Thomas's brother George and his father, Lord Northland, remained loyal, but Thomas would inevitably lose his seat in the 1797 election and sought a junction with the rival Belmore interest.[13] This backdrop of aristocratic pride, electioneering intrigue and poisoned personal relationships complicated Richardson's stance on the militia augmentation.

In May 1795, Montgomery-Moore reported serious physical opposition to recruiting parties in Tyrone from 'hedgemasons, who are no better than Defenders'. By June, at a meeting to oversee the augmentation, this unrest became open defiance over clauses in the Act providing for new-raised men.[14] Vestries could either raise £6 by parish cess for each balloted man or substitute; or, if the quota was not met, pay £6 to the county treasurer for the regiment's stock purse. Under the original Act, counties failing to raise their quota were fined £5 annually per man short. By clause 13 of the amending Act, a special session was to be held no sooner than fourteen and no longer than twenty-nine days after the passage of the Act to determine whether the quota was met or the £6 paid. Failing this, magistrates could direct the treasurer to levy £10 from defaulting parishes for every man short. An initial general balloting meeting was to be followed by two subdivisional meetings, all within a maximum of seventeen days from the passage of the Act. Like other counties, Tyrone did not ballot, but raised volunteers.[15] Clonfeacle had reached its quota of sixteen men, all volunteers, who should theoretically have been exempt. But the regulations were ill-understood and there were rumours that militia recruiting officers were exploiting the confusion and profiteering at public expense.[16] Richardson dug his heels in, stated that his parish would provide no more than 26*s* per man and refused to budge even when threatened with the £10-a-man fine.

As 'a person of information and property' and rector of a parish with an 'irregular' reputation, Richardson's opposition was taken very seriously. Montgomery-Moore recommended that Abercorn thwart the truculent rector 'by levying off him … all penalties'. Abercorn was convinced that electioneering underlay the opposition. Richardson's conduct was 'improper and unworthy … and savours strongly of Tom Knox', and he advised Montgomery-Moore to 'trounce' the 'Doctor'. Montgomery-Moore thought he had his man at a meeting convened to see how far parishes had met their responsibilities, at which Richardson had 'avowed' himself by signing a paper. However, events proved that Richardson had a more accurate reading of the legislation by correctly interpreting clause 13 to mean that over twenty days had elapsed. Montgomery-Moore unsuccessfully sought the opinion of the Attorney General, Arthur Wolfe (Richardson's Trinity friend), then blamed the 'whispered' junction between the Knox and Belmore interests and claimed that Richardson was 'the tool of others', including Lord Northland and another influential landowner, James Verner.[17]

Initially Abercorn believed this, but both men denied any bad intent, and Thomas Knox's brother George cautioned his patron that 'people [i.e. Montgomery-Moore] when they act illegally would … represent those who do not join them as opposing you'.[18] Abercorn explained that his suspicions of Thomas Knox were natural, having previously used him to communicate with Richardson. Dreadfully exposed by conflicting allegiances to his parishioners, Knox and Abercorn, Richardson soothed Abercorn by arguing that he was not politically motivated but that, after the controversial 1793 militia levy, he was pledged to 'my parishioners to stand by them (their own phrase) to oblige me to oppose what I conceived to be a grievance'. This *quid pro quo* rings true. Walter Ricky noted that his parishioners had tried to 'throw every difficulty in his way in respect to his tythes'.[19] Though denying any intention to 'set his opinion up against so respectable an authority as the marquis of Abercorn', Richardson stubbornly justified himself, saying that any levy would cost the county quadruple what it put in the stock purse.[20]

Abercorn's supporters remained divided. John Stewart insisted that Richardson acted 'from the directions' of the Belmore party, but George Knox claimed that Richardson's suggestion that Abercorn should compromise with the protesters would actually 'disconcert the plans of your county opponents'.[21] There the matter lay. It can be understood as an unwanted by-product of Richardson's intermediary position. People would not tell Knox anything, as he was too close to Abercorn, 'while to me who am hardly considered as known to his Lordship people speak their sentiments freely'. Conversely, Abercorn seemed 'to feel uncomfortably towards the people of this country', but Richardson asked Knox to 'set him right', thus ameliorating any elite–plebeian tensions.[22] This productive sensitivity to local sentiment re-emerged in raising the yeomanry in 1796.

During 1795 and 1796 the United Irishmen gained in strength by allying with the Defenders, a Catholic secret society emanating from the sectarian feuding which produced the first Orange lodges, in 1795. In this deteriorating situation some Protestant landowners and magistrates deemed the militia inadequate for internal defence, as southern regiments serving in Ulster were mainly Catholic and suspected of having been infiltrated.[23] Few regulars were available and volunteering had been banned since 1793, as many remaining corps were so radical that they supported the French. This removed a major plank in Protestant defence. Stop-gap loyal associations were mooted as an alternative in 1795, but aborted due to fears about their uncontrollability.[24] The government understood the need for loyal gentry to arm their tenants but, with politics fragmenting between revolutionaries, radicals, reformers and conservative loyalists, how could they create a force immune to political problems and sectarian feuding? Richardson reckoned he had the answer, but it involved Thomas Knox.

Richardson later outlined his remedy in his first published pamphlet, in which he claimed to have devised a yeomanry plan to be used as a template throughout Ireland.[25] He described how, during 1796, he and Thomas Knox had discussed United Irish expansion and the lack of a coordinated loyalist response. Deciding that 'the issue must be decided by the sword' Richardson wrote resolutions for the Tyrone magistracy at the July quarter sessions in Dungannon. These were provisional loyalist offers of service as armed yeomanry, dependent on official acceptance. Those offering would, unlike the Volunteers, accept the command of commissioned officers. Although the grounds of the problem were different from the 1795 militia scenario, it had similar social and political dimensions, and involved the same uneasy trio of 'actors': plebeians, gentry and government. Richardson understood this and so pitched the resolutions 'in such a way as I thought would be acceptable to government, yet would not be rejected by the loyalists, especially if matters were so managed, that it should appear to originate amongst themselves'. Yeomanry were vital in sustaining loyalism and in this 'Dungannon Association' Richardson again mediated between plebeian, gentry and government. When the yeomanry plan was in gestation he had contacted Robert Lindsay, Assistant Barrister for Tyrone, who suggested sending it to the government through the Attorney General, Arthur Wolfe. Richardson did this and got sufficient guarded approval to allow the magistrates to endorse the resolutions. As Knox collected signatures for the 'Association', Richardson embarked on an odyssey to drum up gentry support elsewhere in Ulster.[26] But everything did not go smoothly and his mediating skills were required again.

Despite his family influence at Coleraine, the plan was initially baulked there, and 'schisms' also emerged in Armagh, where Lord Charlemont, once Volunteer commander-in-chief, faced problems from different types of loyalist. Richardson sized up the political landscape and thought the split was 'a freemason business'. The contending groups were old royalists or loyalists, terms he used interchangeably, and 'Republican reformists now disposed to loyalty'. Although the government issued officers' commissions, recommendations came from the ranks and each group had its own nominees. Richardson spoke to all concerned. The 'royalists' heard that intransigence was counterproductive but replied: 'shall we who have always fought the battles of loyalty be superseded ... by the very men who ... debated an hour whether they would take the Oath of Allegiance'. Failing a local solution, Richardson had an ingenious remedy involving his minute understanding of the regulations. The offers of service should proceed in an undifferentiated way; Richardson would personally take them to Dublin Castle, but the government, seeing too many names for one corps, would divide them. This ruse would ensure that any division 'which would be such a triumph to the enemy' was obscured in bureaucracy, but the invidious responsibility for selection transferred from locality to centre.[27] Little wonder that the Speaker of the Irish parliament, John Foster, called the yeomanry scheme 'a task of infinite delicacy'.[28]

Despite Richardson's success in eventually getting the yeomanry scheme off the ground, the momentum of United Irish expansion ran on, and the situation continued to deteriorate. To convert loyalty, a passive and internalised sentiment, into an open and organised movement capable of physically resisting revolution required all points in the triangular model to be in firmly in place and working in conjunction. The immature loyalist organisation of 1796–97 was vulnerable if any of the elements were missing. All hell broke loose when the Tyrone gentry were absent at the 1797 spring assizes.

The northern Protestant outlook is sometimes represented as a 'siege mentality'.[29] By April 1797, Richardson was literally under siege and his 'personal situation ... critical'. Previously he had relied on the physical strength of his Clonfeacle glebe house, the number of its defenders, his own defensive precautions and the proximity of Charlemont Fort. But now huge numbers were becoming passive United Irish supporters by taking the 'oath of secrecy'.[30] Clonfeacle's servants would surrender his arms to save their lives and, while the yeomen might obey orders, 'no numbers can protect the face of a country from an organised banditti who make their next nocturnal attack when they please'. 'Don't tell me', he complained bitterly to Arthur Wolfe, that 'the country is lost if the gentry leave it – be assured all influence is over'. The yeomen were 'a nuisance as they require protection'. Richardson stayed, but only because the army kept the roads open for retreat. His uncharacteristically negative state of mind is apparent in his parting remark to Wolfe, 'you never saw me down before'. His reason for despair was that loyalty had not yet hardened into loyalism, as he had planned, hence his dread of 'yeomanry disaffection'.[31] With the gentry intimidated and the government, in the form of the army, unable to offer protection, the remaining point of the loyalist triangle, plebeian loyalism, was unable to stand alone. There was no relief in the siege of Clonfeacle. Its rector, facing annihilation, fled for the relative safety of Dublin, sending his children to relatives in county Londonderry, where, he glumly remarked, '*they* have not committed the crime of taking tithe'.[32] Richardson may have been down, but he was far from defeated.

The date of his return from Dublin is unknown, but he utilised another contact that united gentry and government to deal with the wavering yeomen. Thomas Knox's younger brother John was a major-general in the army and was responsible for central Ulster.[33] General Knox's response in May 1797 to yeomanry disaffection certainly bears Richardson's, and perhaps Wolfe's, legalistic imprimatur. Knox put a 'test' oath to various yeomanry corps to effectively cancel the secrecy oath by making them swear publicly that they were not United Irishmen.[34] This development, along with a move (discussed below) encouraging Orange admittance into yeomanry corps, ensured that active loyalism not only survived but grew, so that in 1798 the yeomen 'or Orangemen as their opponents thought fit to call them', saved central Ulster.[35]

After the rebellion, Richardson continued to exploit the Knox connection to ensure the yeomanry provided government with the maximum military potential. As part of his plans against possible French invasion, the commander-in-chief, Cornwallis, wanted the force divided into stationary and moveable brigades, but their 'hearths and homes' terms of service militated against strategic flexibility.[36] Richardson and General Knox laid plans for 4,000 hand-picked yeomen to be prepared to leave their districts against an invader until the regular army could assemble.[37] Although this never came to fruition, its implementation would have required the kind of detailed local knowledge that only Richardson and his cohorts had. John Knox died at sea in 1800,[38] but Richardson kept their idea alive. In 1803, after the end of the Peace of Amiens and with invasion apparently imminent, Addington's government hurriedly expanded the United Kingdom's voluntary manpower. Seeing the opportunity, Richardson turned again to his friend Arthur Wolfe, now Lord Kilwarden. They met in Dublin to discuss how Ulster's surfeit of loyal manpower could be used against invasion and Richardson advocated the 1799 plan.[39] Ulster was 'tranquil' and its yeomanry loyal, but there was a control problem which he conceptualised in social and patriotic terms.

Yeomanry captains were often absentees and subaltern officers 'low people' without authority; moreover, a combination of poor direction and intense localism effectively trammelled loyal manpower. 'Alert' yeomen should be organised under good officers and, if they moved to support the regulars, their places were to be filled by loyal citizens. Selection was invidious, sensitivity essential, so 'whatever is done must seem to originate with themselves and thus might be managed'. Knowledge of the dynamics of local loyalism was essential, and Richardson recommended Colonel Trotter. He had the same personal rapport with locals as General Knox once had, yet appeared 'superior to no one'. Kilwarden asked for a summary for the Viceroy, Lord Hardwicke, and Richardson consulted officers and privates before compiling an acceptable form of words. This diplomatically noted that single men offering themselves as 'alerts' for wider service were no less 'prompt' than married men whose families, farms and businesses made them static. Richardson knew Hardwicke would see this and puffed James Verner, who, since 'Tom Knox ... is now lost', was the only person within a twenty-mile radius of Clonfeacle with sufficient weight to drive the measure forward.[40] But his plans were overtaken by events. Civil–military rivalry scuppered anything involving yeomanry, and Kilwarden died in the dust of a Dublin street under the pikes of Emmet's rebels.[41] Richardson advocated the plan to Archbishop Agar and the Chancellor, Lord Redesdale, in 1805 and also during Arthur Wellesley's Chief Secretaryship in late 1807. But he failed to put yeomanry back on the agenda. He admitted to Greenough that Ulster was tranquil, with 38,000 loyal, disciplined yeomen willing to march elsewhere but who 'could not be trusted outside their *own* North'.[42] These yeomanry machinations reveal much about the nature of Irish loyalism and the limits of its impact.

Union distorted the triangular model which had originally worked so well for Irish loyalism. Increased geographical distance attenuated the governmental aspect as parliamentary representation moved to London, while, at the same time, amalgamation of the British and Irish military establishments narrowed room for manoeuvre.[43] The three-card trick had worked best under the Irish parliament, when the combination of high-level government and military contacts with precise and reliable local information resulted in yeomanry formation. Richardson achieved what he did only because, whatever his personal idiosyncrasies, people knew and trusted him. He, in turn, knew and understood them as only a rural rector embroiled in Church, legal and military networks could. Punctiliously a man of his word, he had stood by his parishioners over the militia ballot; yet, through his close friendship with the Attorney

General, he could also relay concerns straight to the heart of government. Richardson's working through Wolfe to achieve governmental sanction for the yeomanry meant that he appeared to have the power to translate and mediate central policy for local consumption. Union altered the dynamics of Irish loyalism by transforming this proximity factor. Despite still being able to call on Ascendancy figures like Kilwarden, Agar and Redesdale, his ideas were ignored. Similarly, processes of negotiation, engagement and special pleading concern the Orange Order. Richardson, though not an Orangeman, was heavily involved. Here again his significance comes from his links with both ground-level loyalism and the Ascendancy elite.

Orangeism and the yeomanry

Richardson had watched commemorations of William III's birthday at Trinity long before the establishment of Orange lodges in Ulster in 1795. However, his influential political linkages and local standing encouraged one of these first lodges to make him privy to their rules. The significance of this cannot be underestimated. Orange lodges were formed by victorious plebeian Protestants in Armagh and Tyrone following the Battle of the Diamond in September 1795, after which the 'Armagh outrages', a systematic persecution of Catholics, occurred. A few landed gentry intervened to try to dissuade the Protestants from violence and a 'loyal association' was attempted by Lord Northland. This would have formalised gentry direction of turbulent plebeians, but the government refused permission, fearing such an association would be hard to control and might worsen the situation.[44] Controlling Orangeism exercised Richardson and, writing later to Greenough, he recalled early 1796, when local combinations were forming against 'the associations of rebels'. Richardson was unenthusiastic about groups mirroring their revolutionary counterparts, but when they showed him their rules he found them unobjectionable. He had neither 'power nor disposition to stop them' and simply recommended that they adhere to their rules, which he forwarded to the Attorney General, who also approved. The men explained that a similar society had begun at the Dyan in county Tyrone and that they were the third such formation.[45] Orange records show that the Dyan received the first warrant and Moy became Loyal Orange Lodge No. 3, so the men Richardson met were members of one of first Orange lodges.[46] We have to rely on Richardson's own retrospective testimony here, given in a plea for English politicians to understand the Ascendancy position coming under hostile scrutiny. Nevertheless, the scenario rings true. Holt

Waring, a local magistrate, received similar Orange rules in July 1796 and also accepted them as signifying law-abiding intent.[47]

The significance of Richardson's experience for penetrating the dynamics of loyalism is twofold. Firstly, the Moy Orangemen were not summoned by Richardson but voluntarily approached him with their rules. Moy lies within Clonfeacle and, as the original Orangemen were Episcopalians,[48] these men would have been his parishioners and probably members of his own congregation, so they were not strangers. As in Richardson's militia pledge to parishioners, this encounter suggests that the structure of power within plebeian loyalism was more horizontally orientated than 'top down'. Secondly, the relationship was not a 'given' and had to be negotiated. Therefore Richardson's securing the Attorney General's approval for the Orange rules must have created the impression in Orange minds that governmental acceptance was within reach on fulfilling certain conditions for keeping the peace. In fact, this was so. The Viceroy, Camden, had no thought of approving of Orange lodges.[49] As we shall see, the government's attitude towards the pragmatic military use of Orangemen softened in 1797 and 1798, though never becoming wholehearted acceptance.[50] The important thing is how people perceived events. By 1798, many Orangemen were convinced of official sanction, assumptions traceable to Richardson's discussions with the Attorney General.

Richardson had an instinctive wariness of any organisation originating outside the elite as being uncontrollable. His reaction to Freemasonry reflected this, even though, as is now known, some were loyalists.[51] Armagh's Masonic committee was loyalist but Tyrone's was mainly radical, due to prominent United Irish Masons like Dr James Reynolds and William Richardson (no relation) of Moy. Richardson's anti-Masonic animus outlasted the rebellion. In 1803 he maintained that any Masonic structure implicitly aided revolutionary principles.[52] Favouring organisations with a control system mirroring the social hierarchy, Richardson saw the yeomanry, as well as being a counter-insurgency force, also as a means to control plebeian loyalists. In the Dungannon Association he had used pre-existing Orange structures without actually naming the supporters as Orangemen, to ensure it had viable popular support.[53] The germ of this particular idea also originated with Wolfe.

At the 1796 Armagh spring assizes, Wolfe had capitally convicted one Protestant and one Catholic for arms raiding, though the Orangeman was later reprieved.[54] Following Richardson's securing of Wolfe's approval of Orange rules, this conviction was intended as a warning to Orangemen, who indeed thereafter 'abstained from further outrages'. Subsequently, when Dungannon Association members offered yeomanry service, the Orangemen signified to neighbouring gentry and to Richardson that they 'would meet no more as a party'.[55] Given this delicate process, Richardson's voice may well have guided Thomas Knox's pen when he told the government that:

> As to the Orange Men we have rather a difficult card to play, they must not be entirely discountenanced, on the contrary, we must in a certain degree uphold them, for, with all their licentiousness, on them we must rely for preservations of our lives and properties should critical times occur. We do not suffer them to parade, but at the same time applaud their loyal professions. I hope I shall be able to manage it that they shall not be lost to the cause of their King and country and at the same time, be kept within due bounds.[56]

The Dungannon Association was unquestionably a major tributary of the yeomanry and Richardson hoped that the affiliation of Orangemen in this government-controlled, gentry-officered force would stop plebeian disorder. Not everyone was convinced and, with Thomas Knox involved, Abercorn was hostile, and said that 'If a single man of mine should enlist with Mr Thomas Knox, who … considers everything merely as the job of the day … I shall feel inclined to give up'.[57] Most Tyrone yeomen lived on the lands of the Dungannon barony and Richardson considered this as the frontier and hoped his rapprochement with Orangemen via the yeomanry had secured that area.[58] North-western Tyrone, where Abercorn's property mainly lay, was different. It bordered on Londonderry and Donegal, maritime counties where a French invasion was feared, as the counties had a substantive United Irish presence. Abercorn's district was vital in the overall securing of Ulster, yet this meant a rigorous security policy involving Orange lodges and yeomanry. Abercorn, who had once promoted the cause Catholic relief, had a social distain for plebeian Orange lodges and still insisted that Catholic disaffection was 'due to the Orange-boy system'.[59] Thus, while Richardson and Knox now believed that Orangeism, handled properly, was the answer, Abercorn saw it as part of the problem.

The key to unlocking Abercorn's reluctance to countenance plebeian Orangeism was Richardson's conviction that it could be subsumed in the government-controlled gentry-officered yeomanry. The grounds for that conviction were near at hand, in the person of General John Knox. He had made Richardson's Clonfeacle house his headquarters. A commissioned officer with a local command and a scion of a Tyrone landed dynasty, he united the government and gentry aspects of loyalism. Initially he brought 'all the English prejudice against Orange' and wanted to stop yeomen wearing orange ribbons on their uniforms, but Richardson soon convinced him that Orangemen were the only guaranteed loyal party and had a strategic utility if absorbed into the yeomanry.[60] Knox accepted this and later reviewed yeomen '*in state* both himself and his horse highly ornamented with Orange ribbons'.[61] Knox and his superior, General Lake, embarked on a rigorous policy by proclaiming districts under the Insurrection Act as exempt from normal law. Knox stirred up tensions, searching for unregistered arms, 'not so much … to succeed to any extent, as to increase the animosity between the Orangemen and the United Irishmen', to secure central Ulster.[62] With *de facto* martial law operative in central Ulster, pressure fell on other areas and in February 1797 Abercorn's stronghold in Strabane Barony was proclaimed.[63] However, tensions between military pragmatism and aristocratic dominance could disrupt the entire project. Intimate with the Knoxes and Abercorn, Richardson became the catalyst of change.

He sent Abercorn three long letters. Historians know of this correspondence but miss its significance. The first letter has been reproduced verbatim for the evidence about sectarian disturbances.[64] However, such information was only part of a polemic to prove that the new military policy would neither encourage plebeian disorder nor compromise Abercorn's political sensibilities, and the letters must be read sequentially. Richardson drew a category distinction between Orangemen, who had no grievances, and previous protest groups. Grievances could be seized on by cynical manipulators to platform their own demands, as the United Irishmen had done. Catholic emancipation was being exploited in this way and Richardson denied that Catholics were inherently disloyal, noting that Clonfeacle's Catholics had stayed clear of the sectarian problems of 1795. The local priest was 'a good humane man and Lord Northland an impartial magistrate'. Local Catholics 'don't trouble themselves' about emancipation, nor were they keen to join the United Irishmen, as most 'never liked, but now abhorred, French politics', and 'nine out of ten of their priests are loyalists'. He had unsuccessfully asked Orangemen to name one Catholic United Irishman, though there were 'plenty in any other description'.[65]

Richardson claimed that yeomanry structures and discipline operated as a kind of behavioural filter. He explained that 'the best of them [the Orangemen] are in the yeomanry', but the 'banditti who robbed in their name, follow their trade without that pretext'. If the Orange name caused problems, Abercorn need not use it as, with loyalty gaining even among those previously disposed to reform, absorbing Orangemen into the yeomanry simplified the political arithmetic: 'our parties are all …

merged into two: *loyalists* and *traitors*'.[66] Abercorn was close to Pitt and Richardson's use of the term 'loyalists' suggests that he envisaged a counter-revolutionary movement, similar to conservative Britons organised with government approval in loyal associations.[67] Some disorder, like the Armagh outrages, could not be ignored, but Richardson advised Abercorn not to believe all he heard about the Orangemen who had stopped the seditious *Northern Star* circulating in Armagh. Richardson introduced a theme which would recur in future efforts to promote the Protestant position: misrepresentation. Although the Armagh outrages had long ceased, revolutionaries continued to publish exaggerated accounts of Orange sectarianism as propaganda.[68]

Richardson's final letter noted that insurgency was not a Catholic problem. The worst counties, Antrim, Down and Derry, were 'the most Protestant part of the north' but had few yeomanry corps, whereas Armagh and Tyrone had numerous loyal Catholics and yeomen. Though martial law would 'excite great complaints', loyalists had nothing to fear and Abercorn would be glad to hear that these 'most desperate' measures were due to governmental hesitation and would have been avoidable if existing laws had been effectively implemented earlier. Having flattered Abercorn politically, Richardson steered him away from the Whig opposition stance which saw martial law as illegal.[69] Whigs mistakenly tried to pacify Ireland by conciliating grievances, but Tyrone's Whig MP, James Stewart, whose 'great object' was popularity, had failed to raise yeomen.[70] General Knox endorsed Richardson's argument and the point was gained. Abercorn accepted that his district, rather than being the most volatile part of Ireland, was now 'the bulwark of the north'.[71]

Abercorn's imprimatur was important. This policy of pragmatically utilising Orange manpower increased in early 1798, an inevitable situation as the growth of lodges outstripped the formation of yeomanry corps.[72] But the marriage of inconvenience had been formalised. By May 1798, General Knox could confidently declare of its offspring that 'the only class of people to take our part are the Orangemen. They are inclined to be licentious, and it requires much difficulty to keep them within bounds. I have found it necessary to encourage that party.'[73] By the 1800s, although Abercorn's agent still dubbed Orangemen 'the very scum of the country', the Marquis had raised a yeomanry legion of 1,200 men, containing only ten token Catholics.[74] Though Abercorn, and presumably Richardson, supported the Union,[75] the measure meant that Ascendancy politics were played out on a wider stage and faced new challenges.

Richardson and loyalism

Though Ireland's executive government continued to be exercised through the Viceroy at Dublin Castle, the transferral (and reduction) of political representation weakened the Protestant political case. Before the Union, amateur militarism and Orangeism had facilitated the tripartite linkage of government, gentry and plebeian. Yet these phenomena were brought about by Ireland's particular exigencies and Irish circumstances could be interpreted in different ways at Westminster. If Irish Protestantism was have its political voice heard in the halls of Westminster, it required effective propagandists.

Loyalism was a prominent theme in propaganda, inviting comparison between Richardson and Sir Richard Musgrave, whose *Memoirs of the different rebellions in Ireland* (1801) presented 1798 as the latest murderous Catholic rising and made his reputation in Britain as the 'primary ideologue' for the ultra-Protestant cause.[76] In 1796 Richardson had unsuccessfully 'offered the assistance of his own pen' for the loyalist cause, but had to confine himself to pseudonymous letters to the *Dublin Journal*.[77] Richardson and Musgrave knew each other and were superficially similar. Both were deeply affected by 1798, supported the Protestant constitution, were involved with the yeomanry and had an intimate knowledge of Orangeism. Beneath this, Richardson was very different, and these differences reveal much about him and more about loyalism. Musgrave himself recognised Richardson's insider knowledge of northern Protestantism and wrote several times for information for his *Memoirs*. Richardson's reaction may be judged from an 1807 letter to Greenough, in which he recalled that: 'Sir Richard Musgrave who, zealous to defend everything Protestant, wrote to me anxious to obtain my admission of the following position that *the Protestant united Irish of the North hearing of the atrocities by Catholics in the south changed sides and became Orange men*'. Richardson refused to comply, turning the tables on Musgrave by saying that the recently formed southern Orange lodges included Protestant United Irishmen who had pragmatically changed sides and their lodges, whose 'avowed motive was defence against the Catholics', were inferior to the 'old loyal Orange' of the north. After several letters 'in which I gave him no comfort as to his new favourites, I at last told him that if he pressed me further I should call out the *Papist Loyalists* – this ended our correspondence'.[78]

This reveals major differences between these two loyalist propagandists on the respective loyalties of Catholics and Presbyterians and on the proper methodology for dealing with the recent past. Musgrave had convinced himself that the northern Presbyterian United Irishmen had 'quitted the papists' and become loyalists in 1798 on finding that Catholics 'were impelled by that sanguinary spirit … ever peculiar to their religion'.[79] Richardson's hostile views on Presbyterian loyalty had modified, but he adduced a different reason from Musgrave. In 1807 he wrote that the northern Presbyterians 'have in the last ten years surprisingly coalesced with the established Church', but the change came not from Musgravian perceptions of 1798 as a Catholic conspiracy to extirpate Protestants, but from political fears about emancipation.[80] Richardson would have refuted Musgrave's wild claim about Tyrone, that its yeomanry amounted to 5,000, of whom 4,200 were Presbyterian and Orange.[81] Moreover, Richardson's assertion that there was, at least in 1798, a north–south tension within loyalism is also worthy of consideration. Indeed, recent work on the early Orange Order has noted a similar demarcation between Ulster and Dublin in institutional and organisational terms.[82] Musgrave was a pivotal figure in the Grand Orange Lodge of Ireland after its establishment in Dublin and became its Grand Treasurer in 1801. But Richardson, as we have seen, had witnessed original Ulster Orangeism evolving in his own parish. The different Ulster attitude to Catholic loyalism that Richardson mentions was conditioned by different circumstances. When the first Orange lodges were formed, the threat came from Presbyterian United Irishmen allied to the Defenders, not, as Musgrave would have had it, from a primarily Catholic insurrectionary force. Richardson believed that 'the Belfast United Irishmen were … all Protestants, hence they made little or rather no impression upon the Catholics', who, unless led astray by demagogues, were naturally opposed to French-style republicanism.[83] There is independent evidence from 1797 of Catholic loyalty being endorsed by some Orangemen.[84] Historians have queried the extent and genuineness of Catholic loyalty, but it is significant that Richardson made these claims at all.[85] The important point is that Richardson's refutation of Musgrave's Catholic conspiracy theory shows that loyalism, like its United Irish counterpart, was not homogenous and had a distinctive and more nuanced northern strain. Accuracy of information aside, Richardson had profound differences with Musgrave over the construction of knowledge itself.

Musgrave's inaccurate statement about Presbyterians in Tyrone reflects the fact that Richardson, the person best qualified to provide information, had refused to do so. Another Musgravian assertion involving the yeomanry brought Richardson's objections into the open. Musgrave's *Memoirs* stated that the yeomanry originated in Dublin under Lord Camden's authority and was modelled on the English yeoman cavalry. Richardson's 1801 yeomanry pamphlet vigorously refuted that claim, a denial which even reached William Drennan's sister, Martha McTier, who heard that 'he [Richardson] contradicts Sir Richard Musgrave's statements in regard to *that* fact'.[86] Richardson's political analysis followed the same methodology as he used for natural history, basing his arguments on hard facts.[87] Musgrave's culpability was a flawed methodology which contravened empirical standards of observable evidence. He started with a theory – Catholic conspiracy – and forced the facts to conform to it. Regarding the claim about the yeomanry, Richardson stated that 'not one of these positions of SIR RICHARD is founded on fact'. As a true empiricist he challenged Musgrave, as he would later challenge Malthus, to 'produce his authorities'; otherwise, he would not be listened to.[88] Richardson's denial to Musgrave of empirical evidence about Tyrone therefore smacks of intellectual superiority as well as personal dislike. For Musgrave, Catholics were naturally predisposed to rebellion and massacre; for Richardson, they were innocents manipulated by others who misrepresented their grievances.

Though Richardson and Musgrave both came to oppose Catholic emancipation, their stance was formulated on different premises and was based on different chronological periods. In 1792, Musgrave had supported the Dublin Corporation's 'Protestant Ascendancy' resolutions using the historical justification that Catholic treachery and disloyalty were 'timeless'.[89] At this same period Richardson reflected Abercorn's views: 'upon the great question of Emancipation (*then* properly applied) I was a warm advocate, and for a long time [afterwards]'.[90] Emancipation, originally intended to accompany Union, remained a dead letter until 1804, when agitation recommenced.[91] By 1807, Richardson had dropped his pro-Catholic line and noted that 'their conduct satisfied me that the small minority who *then* opposed were right'.[92] Richardson's change on this issue reflects a wider shift within Protestant opinion which saw the neo-conservative position on emancipation move to the mainstream. If Musgrave is seen as a trend-setter anticipating wider thinking, he appears as prophetic; yet Richardson's alteration came not from a past-orientated conspiracy theory, but from present-based empirical facts. Other northern magnates knew this. In 1804, when Earl Macartney wanted Ulster gentry views on the renewed emancipation agitation and proposals for state payment of Catholic clergy, he asked Richardson, not Musgrave.

These discussions are revealing about shifts in Richardson's thinking. They discussed the possibility of a state payment for the Catholic clergy, similar to the *regium donum* Presbyterian ministers received. This was considered during the Catholic relief debates in 1793, but rejected by the Catholic hierarchy, who feared, correctly, that the state wanted a say in ecclesiastical nominations.[93] Richardson said that the payment of the *regium donum* was a mistake as ministers of that denomination had been involved in the 1798 rising. Concessions had failed to achieve conciliation and fuelled further demands. Catholic loyalty had disappeared as they now looked to Bonaparte, and Maynooth was 'a centre of virulence' against the state. Englishmen should grasp what Irish Protestants knew, that '*power* is their real object', directed at 'the recovery of their ancient property and church establishment'. Emancipation, in fact, was a fig leaf for Catholic political ascendancy, but for the English 'the question is almost a speculative one, with us everything is at stake'.[94]

Richardson was still optimistic about Catholics at parish level and saw the real danger lying with the hierarchy, whom he called 'the enemy'. State payment was a possible solution, as it would give government control, but it should be 'so modelled that *we* derive advantage from it, and the enemy none'. If payment separated the parish clergy from their superiors and made them grateful to government, it certainly was desirable, but risky. Without 'dextrous management', the Catholic Church structure would produce 'an increase of wealth, power and patronage to their bishops, that is to our inveterate enemies'. He advocated a Machiavellian subterfuge resembling his yeomanry schemes to force the parish clergy, 'as *indirectly* as we can', to teach their flocks '*loyalty* and *peace*'. The authorities must 'make the business transact itself', and thus payment should be neither a state salary nor a bonus. Rather, depending on individual priests' conduct, it should be at the discretion of the '*resident Protestant gentry*' and the state. The intent would be obscured, as 'the importance of a *resident* clergy' could be given 'as an ostensible reason'.[95] He had subtly sounded Clonfeacle's Catholic priest, Father Devlin, whom he had known for twenty-five years, which had 'enabled me to measure him'. Devlin was 'a humane and … good man, anxious to preserve the peace of the country'. But he was also 'a bigoted Catholic' who accepted the view of Catholic Vicar-General, Dr Conwell, view that payment would diminish the hierarchy's influence over their people. Richardson 'hinted a few difficulties to try him', but Devlin was 'perfectly master of the question'. In any case, for Richardson at this stage, the Catholic question was 'dormant', as 'their importance (as was foreseen) is lost in the mass of Empire'. The danger was Irish Whigs misrepresenting the issue, as 'they care as little for the success of Catholic claims as you or I, but they are very anxious to hold their posts as leaders'.[96]

By the time he wrote to Greenough in 1807, the pro-Catholic 'Talents' ministry had collapsed over emancipation and this gave Richardson more empirical evidence for his opposition. He had experienced the courting of Catholics by Irish Whig supporters of the Talents. Though the government had changed, the danger remained, and if Richardson could convince English Whigs that Ireland, under the present political disposition, was an asset to the United Kingdom, so much the better. Better than the behaviour of the Tyrone magistrate, Richard Wilson.

Wilson was originally from Hertfordshire and had been MP for Barnstable but, on losing his seat, came to Ireland in 1802 to live at his mother's property, Oona Lodge, near Clonfeacle. He was a spendthrift who left debts, sexual scandals and a broken marriage behind him. In parliament he voted with the Foxite Whigs and, as a Tyrone magistrate, assumed the role of protector of the Catholics, alleging that they could not get justice from Orange magistrates. This angered the Protestant gentry, especially as Wilson published his complaints in pamphlets.[97] He maintained the attack after the Talents fell, dedicating a pamphlet to the new Viceroy, Richmond, about 'murders and robberies committed ... upon the Roman Catholics by a banditti calling themselves Orangemen'. Richmond planned to tour Ulster and meet its gentry, but Wilson warned 'I see no one name but what belongs to the avowed patrons of Orangeism ... and not friends to the Roman Catholics'. Specifically he alleged that James Verner and his sons winked at the burning of a Catholic's house by Orange yeomen and helped them escape prosecution.[98]

Richmond suspected that Wilson 'meant to do mischief', as a London edition of William Parnell's influential emancipationist pamphlet included a section on Catholic exclusion from juries.[99] Richardson, fearing such accusations could be exploited, unleashed a volley of refutation and counter-propaganda.[100] The house burning, he said, was personal, not sectarian. Protestants and Catholics may not have 'loved' each other, but saw the benefits of peace and any agitation, like inflammatory petitions, was artificially stimulated. Allegations of partial justice were disproved by the district's freedom from sectarian disputes for the last ten years. Richardson enclosed affidavits proving that some propertied Catholics served on juries and their community was grateful to Verner, blackened by Wilson as a Catholic persecutor, for granting ground for a new chapel.[101] Wilson was a 'bankrupt' who tried to improve his position by inflaming sectarian tensions and, having failed, became a 'trading justice', a point Richardson substantiated by eighteen affidavits complaining about fines for petty matters.[102] Though the lines of communication are irrecoverable, it may not be coincidental that the new under-secretary, James Traill, said that Verner's leasing policy to Catholics disproved Wilson's accusations.[103]

Tithes were another area of Ascendancy vulnerability, as the Whigs saw reform as a way to pacify Ireland. Richardson assured Greenough that Ireland's various protest movements never had tithes as a primary grievance. Catholics did not object to it; indeed, Clonfeacle's priest had assured him in 1784 that his flock would not join any anti-tithe combination. British parliamentarians, like Greenough, therefore should not rush headlong into reform due to a 'forced clamour' by 'the lately ousted party'. Here, as elsewhere, Richardson voiced 'rooted unconquerable antipathy' to the Whigs. 'Do you belong to the Talents?', he asked his friend; if so, 'any political information is unnecessary, as they understand everything, and they alone'. Knowing the tithe issue would be raised by Henry Grattan, Richardson provided local information that anti-tithe meetings excited little notice. Anglican clergymen were less concerned about income reduction than about any new commutation system which they feared would disadvantage those least able to pay. Again, Richardson presented his information from a northern angle. In the early eighteenth century, the Irish parliament exempted pasturage from the tithe of 'agistment', but that, he said, should now be revoked and the income used to compensate clergy, in return for exempting flax and potatoes, mainstays of the poor, thus privileging Ulster's small tillage weaver-farmers over southern graziers. The process worked two ways. Richardson supplied information to his uncommitted Westminster contact but used him as a source of political intelligence, asking Greenough to 'communicate to me *any plan* in agitation, [that] I may comment upon it'.[104]

Greenough also wanted to know about conflict between sectarian parties. Richardson ironically complimented him on being the first Englishman who had actually bothered to ask, though many thought they knew better than the Irish themselves. But he responded nonetheless. Regarding Catholics, 'you think you have some idea, probably very erroneous', while 'of *Orange*, you know absolutely nothing, or worse than nothing'. Richardson denied being a 'prejudiced Orange man' and explained how he had become a special constable during the Armagh outrages.[105] A further question prompted a genealogy of disorder broadly similar to that given to Abercorn. But the justification to an Irish magnate in 1797 was different in degree from the explanation to an English MP in 1807. Richardson made the same distinction between law-abiding Orange lodges and 'racking' by 'the dregs and refuse of Protestants, mostly journeymen weavers', who 'for some time were called Orange men, but did not assume the name'. Pressed about why they were sworn to William III and not George III, he claimed the Orangemen's enemies were those of King George (the United Irishmen and the French), not those of William of Orange (Irish Catholics). In any case, the government need not worry, as the Orangemen were 'so merged in the yeomanry, which has completely fulfilled all their objects', that they appeared only on 12

July. This selected evidence was designed for a particular purpose, which emphasised Orange war patriotism, which would have been recognisable in Britain, but downplayed sectarianism. Yet Richardson's instincts about Westminster being a cold house for Orangeism were correct. His question 'would you stop their parading and festivities on that day?' anticipated later legislation suppressing parades.[106]

In January 1809 Greenough solicited Richardson's views on the veto over Catholic ecclesiastical appointments, a topic which had recently been addressed in several pamphlets. Richardson told him that he had once supported Catholic relief but 'in the course of 36 years have acquired so thorough a knowledge of them that if a *real* grievance still remained upon them I would refuse to redress it', as even that was subject to Whig misrepresentation. The context was the debate on the 1808 emancipation petition in which the Prime Minister, Lord Grenville, incorrectly claimed that he was authorised by the Irish bishops' English agent, John Milner, to say that the hierarchy would accept a government veto. This left Grenville and his supporters in a difficult position.[107] Ascendancy Protestants like Richardson relished their embarrassment. '*We*', he said expansively, 'expect much sport in parliament ... and long to see how Grattan and Ponsonby extricate themselves'. Although Greenough had not spoken in parliament, Richardson suggested an opening speech advocating a similar parochial clergy payment scheme as discussed with Macartney in 1804. Greenough was advised to treat the question objectively, Richardson noting how the *regium donum* was a good precedent, and affect concern about poverty by ending dues to priests. Behind this, however, 'though without betraying an intention of the kind', the issue was a political litmus test. If accepted, the government gained a handle on the 'working clergy'; if rejected, it showed the Catholics were 'disloyal'. Greenough never made any speech before relinquishing his Gatton seat in 1812, but the tenor of Richardson's correspondence shows how Union had compromised Ascendancy.[108] The position on militarism, Orangeism and Protestant politics, originally created by Irish methods and formulated around Irish issues, had to be re-configured and re-defended. What, in conclusion, does this reveal about Richardson and more widely about Irish loyalism?

Loyalism and Protestant Ascendancy

Richardson emerges to be a promoter of the Ascendancy cause but, unlike Musgrave, he preferred to remain in the background. Whatever the verdict of hindsight, however, less visibility does not equate to diminished contemporary significance. Richardson's advocacy falls into two very different spheres, broadly before and after Union. Before 1800, Richardson's promotion was direct and by deed, whereas afterwards it was indirect and more like classic propaganda, as ideas and arguments were privileged over actions, with Richardson using Greenough as his main surrogate. In the former period, Richardson operated as an intermediary, with demonstrable connections to government and the upper echelons of the judiciary through his Trinity friendships and close relationships to key military personnel. His influential position in Tyrone made him indispensible to the gentry and his role as a rural rector brought intimate contact with plebeian parishioners. From 1796, Richardson aimed to steady loyal sentiment and construct a sustainable loyalist counter-revolutionary movement under government and gentry control. His efforts to achieve this through amateur militarism and loyal associationalism show that loyalism, as he envisaged it, needed all ingredients cooperating together. Although he consistently advocated elite control, his involvement in loyalism's construction shows that delicate negotiation with plebeians was essential. This exploitation of local knowledge to negotiate terms of service with Orange yeomen mirrors the view historians now take about English loyalism empowering its adherents.[109]

Richardson's differences with Musgrave over the respective loyalty of Presbyterians and Catholics and his claims for the superiority of the northern loyalism he helped construct over its southern counterpart reveals a distinct Ulster strain. This has largely disappeared from historical view owing to the moves for Protestant political unity during the O'Connell era. Indeed, Richardson himself was told to stop emphasising the singularity of northern loyalism in the interests of Protestant unity. Richardson did not base his opposition to emancipation on emotive massacre fears or even on the rebellion, but on pragmatic political concerns about the power of the Catholic hierarchy and worries that Irish Whigs would seize on the issue as a step on the political ladder. Northern Anglicans, even after 1798, conceived the threat they faced as coming from other Protestants as much as from Catholics.

Richardson knew his own value to the Protestant cause. When Macartney asked for the gentry position on emancipation Richardson gave his credentials, noting that 'for the last 25 years, a constant resident in Tyrone or Antrim, I took an active part in the busy scenes that were passing around me [and] of course cannot be ignorant of the state of parties and the deposition of the people of the country'.[110] A careful reading of Richardson's letters after Union reveals how the Ascendancy position was weakened and why. Richardson's confident use of the pronoun 'we' in giving the Protestant position suggests that he represented a wider constituency and was not a voice crying in the wilderness. If he was not exactly in the political wilderness, Richardson's activities from 1804 to 1810 show that the Ascendancy cause did suffer from isolation. London was less accessible than Dublin, but underlying all was the unpalatable reality of the loss of communication networks, often enmeshed in family and educational connections, which had brought College Green into the purview of provincial Protestantism. The shock is palpable in Richardson's heartfelt complaint about how crucial political pamphlets, like Parnell's on the Catholic question, 'had not reached us here'. Communication in either direction could no longer be taken for granted: it had to be worked at. Richardson tried hard to restore the *status quo ante* through his English connections, acting as a conduit for his college friend Dr Thomas Elrington's pamphlet on tithes. Ignorance of Ireland was another potentially debilitating factor and Richardson complained that Englishmen did not understand the disposition of religious groups in Ireland and should realise that the Catholic issue was not about religion, but power for the upper orders and property for the lower.[111]

One ubiquitous factor in all Richardson's political considerations was the fear of misrepresentation. The Catholic campaign for emancipation and the Irish Whigs misrepresented plebeian grievances, the United Irishmen misrepresented Orange disorder, Musgrave misrepresented loyalism and Richard Wilson misrepresented the Tyrone magistracy. Arguably there was a connection between this fear and the threats facing the Ascendancy, a danger exacerbated by distance after Union. The Protestant position was certainly under threat, both personally in Richardson's virtual physical besiegement in 1797 and more generally in what must have seemed like a political siege during the Talents administration. Below the level of consciousness, there may have been some subterranean cultural reason for Richardson's fear of misrepresentation arising from the Protestant experience in Ireland. However, a more convincing explanation lies in Richardson's insistence on basing his analyses only what he saw as the demonstrable facts, with no allowance for subjectivity. This approach arose from his intellectual grounding in the empirical methods of Francis Bacon and was to come to the fore in his writings on geology.

Notes

1 I. McBride, *Eighteenth-century Ireland* (Dublin, 2009), p. 2.
2 A. Blackstock, *Loyalism in Ireland, 1789–1829* (Woodbridge, 2007), pp. 13–15.

3 K. Whelan, *The tree of liberty: radicalism, Catholicism and the construction of Irish identity* (Cork, 1996), pp. 42–3, 123, 142–3.

4 D. Eastwood, 'E. P. Thompson, Britain and the French Revolution', *History Workshop Journal*, 39 (spring 1995), 81–2.

5 Blackstock, *Loyalism*, pp. 267–71.

6 Barnard, *A new anatomy of Ireland*; Connolly, *Religion, law and power*; A. P. W. Malcomson, *John Foster, the politics of improvement and prosperity* (Dublin, 2011); A. P. W. Malcomson, *John Foster: the politics of the Anglo-Irish ascendancy* (Oxford, 1978); Malcomson, *Archbishop Charles Agar*; J. Hill, *From patriots to unionists: Dublin civic politics and Irish Protestant patriotism, 1660–1840* (Oxford, 1997); J. Kelly, *Sir Richard Musgrave, 1746–1818: ultra-Protestant ideologue* (Dublin, 2009); J. Kelly, *Sir Edward Newenham MP, 1734–1814: defender of the Protestant constitution* (Dublin, 2004).

7 R. G. Thorne (ed.), *The House of Commons, 1790–1820* (London, 1986), vol. 3, p. 172; John Wyatt, 'Greenough, George Bellas (1778–1855)', *Oxford dictionary of national biography* (Oxford, 2004), online edition, 2004, www.oxforddnb.com/view/article/11432, accessed 19 August 2011; UCL, Greenough's journal, 7/13.

8 Blackstock, *An Ascendancy army: the Irish yeomanry, 1796–1834* (Dublin, 1998), pp. 41–3, 53.

9 S. J. Connolly, *Divided kingdom: Ireland, 1630–1800* (Oxford, 2008), pp. 403–5; M. R. O'Connell, *Irish politics and social conflict in the age of the American Revolution* (Philadelphia, 1965), p. 91.

10 I. Nelson, *The Irish militia, 1793–1802: Ireland's forgotten army* (Dublin, 2007), pp. 49, 68–9.

11 Nelson, *Irish militia*, p. 54.

12 A. P. W. Malcomson, 'A lost natural leader: John James Hamilton first Marquis of Abercorn', *Proceedings of the Royal Irish Academy*, vol. 88, C, no. 4, pp. 66, 68, 73; Johnston-Liik, *Irish parliament*, vol. 4, pp. 40–1, 48–9, vol. 2, pp. 333–39, 380; Thorne (ed.), *The House of Commons*, vol. 5, p. 349; R. Murray to 'Dear Will', 18 June 1791, G. Hall to Richardson, 18 June 1791, PRONI, Abercorn papers, D623/A/133/18, 19.

13 Johnston-Liik, *Irish parliament*, vol. 5, pp. 48–9, 126–7, 284–5; Thorne (ed.), *House of Commons*, vol. 4, p. 349.

14 Montgomery-Moore to Abercorn, 23 May 1795, Captain Stewart to Montgomery-Moore, 8 June 1795, PRONI, Abercorn papers, D623/A/147/21, 25.

15 Sir Henry McAnally, *The Irish militia, 1793–1816* (Dublin and London, 1949), pp. 71–7; Abercorn to John Stewart, 19 June 1795, PRONI, Abercorn papers, D623/A/79/105; 35 Geo. III c.8, 'An Act to explain and amend 33 Geo III c.22'.

16 Captain Stewart to Montgomery-Moore, 8 June 1795, John Stewart to Abercorn, 3 July 1795, PRONI, Abercorn papers, D623/A/139/25, 26.

17 Montgomery-Moore to Abercorn, 2, 13 June, 19, 24 July 1795, PRONI, Abercorn papers, D623/A/147/24, 28, 29, D623/A/148/14; Abercorn to Montgomery-Moore, 19, 21 June 1795, PRONI, Abercorn papers, D623/A/79/106, 107; Abercorn to Camden, 16 June 1795, PRONI, Abercorn papers, D623/A/79/102.

18 Abercorn to George Knox, 2 August; Lord Northland to George Knox, 12 August; George Knox to Abercorn, 13 August 1795, Abercorn to James Verner, 3 September 1795, PRONI Abercorn Papers, D623/A/79/116, 121; D623/A/137/40, 41, 42.

19 Abercorn to G. Knox, 3 September 1795, Richardson to G. Knox, 17 September

1795, PRONI, Abercorn papers, D623/A/79/122, D623/A/137/44; Ricky, *Extraordinary conduct*, pp. 24–7.
20 Richardson to George Knox, 17 September 1795, PRONI, Abercorn papers, D623/A/137/44.
21 G. Knox to Abercorn, 23 September 1795, John Stewart to Abercorn, 29 September 1795, PRONI, Abercorn papers, D623/A/137/46, D623/A/139/29.
22 Richardson to G. Knox, 17 September 1795, PRONI, Abercorn papers, D623/A/137/44.
23 Nelson, *Irish militia*, pp. 150–1.
24 Blackstock, *An Ascendancy army*, pp. 57, 50–1; Blackstock, *Loyalism*, pp. 61, 63.
25 W. Richardson, *History of the origins of the Irish yeomanry* (Dublin, 1801).
26 Richardson, *Yeomanry*, pp. 15, 20–1.
27 Richardson to Arthur Wolfe, 26 September 1796, NAI, Rebellion papers, 620/25/118.
28 John Foster to Camden, 2 September 1796, Kent Archives Office, Pratt papers, U840/0194/3–4.
29 I. McBride, *The siege of Derry in Ulster Protestant mythology* (Dublin, 1997), p. 10.
30 Richardson to Wolfe, 8 April 1797, NAI, Rebellion papers, 620/29/200.
31 Richardson to Wolfe, 8, 9 April 1797, NAI, Rebellion papers, 620/29/200, 204; J. Knox to Pelham, BL, Pelham papers, add mss 33103, fols 351–2.
32 Camden to Portland, 15 April 1797, PRO, Home Office 100 series, HO100/69, fol. 195; Ricky, *Extraordinary conduct*; Richardson to Wolfe, 9 April 1797, NAI, Rebellion papers, 620/29/202.
33 Blackstock, *Loyalism*, p. 72.
34 J. Knox to Pelham, 17 May, 6 June 1797, BL, Pehlam papers, add mss 33104, fols 77–8, 192–4.
35 Richardson, *Yeomanry*, p. 42.
36 Blackstock, *Ascendancy army*, pp. 162–3; Blackstock, *Loyalism*, pp. 123–4.
37 Richardson, *Yeomanry*, pp. 44–8.
38 Johnston-Liik, *Irish parliament*, vol. 5, p. 44.
39 J. Cookson, *The British armed nation* (Oxford, 1997), pp. 82, 95; G. Cousins, *The defenders: a history of the British volunteers* (London, 1968), pp. 84–5; Richardson to Kilwarden, 2 May 1803, BL, Hardwicke papers, add mss 35739, fols 3–6.
40 Richardson to Kilwarden, 2 May 1803, BL, Hardwicke papers, add mss 35739, fols 3–6; Richardson to Kilwarden, 8 June 1803, BL, Hardwicke papers, add mss 35750, fol. 253.
41 A. Blackstock, 'The union and the military, 1810–c.1830', *Transactions of the Royal Historical Society*, 6th series, vol. 10 (2000), pp. 338–9.
42 Richardson to Archbishop Agar, 20 November 1805, PRONI, Agar papers, T3719/C38/23; T. Bartlett, *The fall and rise of the Irish nation: the Catholic question, 1690–1830* (Dublin, 1992), p. 277; Castlereagh to Wellesley, 28 December 1807, in Wellington, *The supplementary despatches, letters and memoranda of Arthur Wellesley, first Duke of Wellington* (London, 1860), vol. 5, pp. 279–83; Richardson to Greenough, 20 December 1807, 10 January 1808, UCL, Additional Greenough papers, 1403, 1410.
43 Blackstock, 'The union and the military', p. 333.
44 H. Senior, *Orangeism in Ireland and Britain, 1795–1836* (London and Toronto, 1966), pp. 18, 29–31.

45 Richardson to Greenough, 14 January 1808, UCL Additional Greenough papers, 1412.
46 Grand Orange Lodge of Ireland, Education Committee, *The formation of the Orange Order, 1795–1798* (Belfast, 1994), p. 62.
47 Waring to Cooke, 4, 23 July 1796, NAI, Rebellion papers, 620/24/11, 46.
48 Senior, *Orangeism*, p. 18.
49 Camden to Portland, 22 January 1796, PRO, Home Office 100 series, 62, fols 15–20.
50 Blackstock, *Ascendancy army*, p. 96.
51 Petri Mirala, *Freemasonry in Ulster, 1733–1813* (Dublin, 2007), pp. 127, 131–2, 220, 232–4.
52 Richardson to Kilwarden, 2 May 1803, BL, Hardwicke papers, add mss 35739, fols 3–6.
53 Blackstock, *Ascendancy army*, pp. 61–2.
54 Senior, *Orangeism*, p. 37.
55 Richardson to Abercorn, 22 February 1797, PRONI, Abercorn papers, T2541/IB3/6/5.
56 T. Knox to Cooke, 13 August 1796, NAI, Rebellion papers, 620/24/106.
57 Abercorn to Stewart, n.d. [?4] March 1797, PRONI, Abercorn papers, D623/A/80/40.
58 Richardson to A. Wolfe, 2 November 1796, NAI, Rebellion papers, 620/33/10.
59 T. Bartlett, *Fall and rise*, pp. 159–61; Abercorn to Stewart, 2 February 1797, PRONI, Abercorn papers, D623/A/80/39.
60 Richardson to Greenough, 14 January 1808, UCL, Additional Greenough papers, 1412.
61 J. Knox to Pelham, 19 April 1797, BL, Pelham papers, add mss 33103, fols 379–80; Lake to J. Knox, 19 April 1797, NLI, Lake MSS, MS56, fols 53, 3; Richardson to Greenough, 14 January 1808, UCL, Additional Greenough papers, 1412.
62 J. Knox to Pelham, 19 March 1797, BL, Pelham papers, add mss 33103, fol. 263.
63 An account of all the parishes proclaimed [n.d. 1797], PRO, Home Office 100 series, HO100/79, fol. 346; W. E. H. Lecky, *Ireland in the eighteenth century* (Cabinet edition, 4 vols, London, 1913), vol. 3, pp. 19–20.
64 D. Miller, *Peep O'Day boys and defenders* (Belfast, 1990), pp. 131–7.
65 Richardson to Abercorn, 14, 22 February 1797, PRONI, Abercorn papers, T2541/IB3/6/3, T2541/IB3/6/5.
66 Richardson, *Yeomanry*, pp. 21–7; for discussion of the evolution of the yeomanry see Blackstock, *Ascendancy army*, chapter 3; Richardson to Abercorn, 22 February 1797, PRONI, Abercorn papers, T2541/IB3/6/5.
67 Malcomson, 'A lost natural leader', pp. 67–9; F. O'Gorman, 'English loyalism re-visited', in A. Blackstock and E. Magennis (eds), *Politics and political culture in Britain and Ireland* (Belfast, 2007), p. 227.
68 A. Blackstock, 'Loyal crowds in mid-Ulster, 1795–96', in P. J. Jupp and E. Magennis (eds), *Crowds in Ireland, c.1720–1920* (Basingstoke and New York, 2000), p. 95; Richardson to Abercorn, 14, 22 February 1797, PRONI, Abercorn papers, T2541/IB3/6/3, 5.
69 Lecky, *Ireland*, vol. 4, pp. 20–1.
70 Richardson to Abercorn, 28 March 1797, PRONI, Abercorn papers, T2541/IB3/6/11.

71 J. Knox to Abercorn, 25 February 1797; Abercorn to Richardson, 17 March 1797, PRONI, Abercorn papers, D623/A/156/6, D623/A/80/41.
72 Blackstock, *Ascendancy army*, pp. 94–5; Blackstock, *Loyalism*, pp. 80–1.
73 J. Knox to Abercorn, 11 May 1797, PRONI, Abercorn papers, D623/A/156/14.
74 Hamilton to Abercorn, 24 April, 7 July 1802, Abercorn to Littlehales, 15 August 1803, PRONI, Abercorn papers, D623/A/94/15, 23, D623/A/81/64.
75 Abercorn to Castlereagh, 4 July 1799, PRONI, Abercorn papers, D623/A/80/126.
76 Kelly, *Sir Richard Musgrave*, pp. 127, 151, 247.
77 Richardson to Greenough, 3 January 1808, UCL, Additional Greenough papers, 1407.
78 Richardson to Greenough, 27 January 1807, UCL, Additional Greenough papers, 1417.
79 Musgrave to Lenox-Conyingham, 27 April 1799, PRONI, Lenox-Conyingham papers, D1449/12/292.
80 Richardson to Greenough, n.d. [December 1807], UCL, Additional Greenough papers, 1401.
81 R. Musgrave, *Memoirs of the different rebellions in Ireland* (Dublin, 1801), p. 194.
82 Kelly, *Sir Richard Musgrave*, p. 154; James Wilson, 'Orangeism in 1798', in T. Bartlett, D. Dickson, D. Keogh and K. Whelan (eds), *1798: a bicentenary perspective* (Dublin, 2003), pp. 345–62.
83 Richardson to Greenough, 14 January 1808, UCL, Additional Greenough papers, 1412.
84 Blackstock, *Loyalism*, pp. 82–9.
85 D. Keogh, *The French disease: the Catholic Church and radicalism in Ireland, 1790–1800* (Dublin, 1993), pp. 167–8.
86 Martha McTier to William Drennan, n.d. [1801], in J. Agnew and M. Luddy (eds), *The Drennan–McTier letters, 1802–1819* (Dublin, 1999), vol. 3, pp. 7–9; Richardson, *Yeomanry*, pp. 8–9.
87 Richardson to Greenough, 24 January, UCL, Additional Greenough papers, 1416.
88 Richardson, *Yeomanry*, pp. 8–9.
89 Kelly, *Sir Richard Musgrave*, pp. 14–15; Hill, *From patriots to unionists*, p. 221.
90 Richardson to Greenough, 3 December 1807, 27 January 1808, UCL, Additional Greenough papers, 1402, 1417.
91 Bartlett, *Fall and rise*, p. 280.
92 Richardson to Greenough, 3 December 1807, 27 January 1808, UCL, Additional Greenough papers, 1402, 1417.
93 Bartlett, *Fall and rise*, pp. 189–90.
94 Richardson to Macartney, 18 August 1804, PRONI, Macartney papers, D572/14/20.
95 Richardson to Macartney, 18 August 1804, PRONI, Macartney papers, D572/14/20.
96 Richardson to Macartney, 5, 8 December 1804, PRONI, Macartney papers, D572/14/17–18; Bishop Reilly to Dr Conwell, 11 August 1805, Tomas O'Fiaich Library, Armagh, Conwell papers.
97 Thorne (ed.), *History of parliament*, vol. 5, pp. 602–3; Bedford to Spencer, 29 May 1806, PRO, Home Office 100 series, 135, fol. 226.
98 R. Wilson, *A narrative of the various murders and robberies committed … upon the Roman Catholics by a banditti calling themselves Orangemen* (Dublin, 1808), pp. v, 82, 86–7.

99 Richmond to Wellesley, 22 January 1808, Southampton University Library, Wellington papers, WP1/188/41; W. Parnell, *An historical apology for the Irish Catholics* (Dublin, 1807, London, 1808).

100 Bedford to Spencer, Elliott to Bedford, 29 May 1806, PRO, Home Office 100 series, 135, fols 226–7; Richardson to Greenough, 27 December 1807, 13 January 1808, UCL, Additional Greenough papers, 1405, 1413.

101 Richardson to Greenough, 3, 6, 7, 10, 14, 17 January, 2, 3, 6 February 1808, UCL, Additional Greenough papers, 1407, 1408, 1409, 1410, 1413, 1414, 1418, 1419, 1420.

102 Richardson to Greenough, 24, 27 December 1807, 13 January 1808, UCL, Additional Greenough papers, 1404, 1405; Richardson, *Yeomanry*, p. 37; see also John Anketell, *A letter to Rev. Henry Conwell D.D. popish priest of Drumglass and Killyman relative to a pamphlet lately published by Richard Wilson Esq.* (Dublin, 1807).

103 Traill to Wellesley, 26 January 1808, Southampton University Library, Wellington papers, WP1/188/56.

104 Richardson to Greenough, n.d. [December 1807], 20, 27 December 1807, 11 January 1808, UCL, Additional Greenough papers, 1401, 1403, 1405, 1411.

105 Richardson to Greenough, 20, 29 December 1807, UCL, Additional Greenough papers, 1403, 1406–7.

106 Richardson to Greenough 14, 27 January 1808, UCL, Additional Greenough papers, 1412, 1417.

107 Bartlett, *Fall and rise*, pp. 291–2.

108 Richardson to Greenough 2 January 1809, UCL, Additional Greenough papers, 1422; Thorne (ed.), *History of Parliament*, vol. 3, p. 172.

109 Eastwood, 'E. P. Thompson', p. 83.

110 Richardson to Macartney, 18 August 1804, PRONI, Macartney papers, D572/14/20.

111 Richardson to Greenough, 24, 27 January, 3 February 1808, UCL, Additional Greenough papers, 1416, 1417, 1419.

3

'The scattered remnants of a diminished world':[1] William Richardson and geology

Mephistopheles, in Goethe's *Faust*, arriving on a jagged mountain top, recognises the place: 'I know it well: this used to be the very floor of Hell, before God tumbled Satan from his throne, and turned all of Creation upside down'. The intellectually arrogant Faust denies he aims higher still, as 'the earth still gives ample scope for what I have in mind'. But Mephistopheles asks, is this not just 'a lust for glory?', echoing the pronouncement of the seventeenth-century natural philosopher Francis Bacon that 'it was from a lust of power that the angel fell, from a lust of knowledge man fell'.[2] Goethe's correlation of geological turmoil with biblical evil had resonance in the supposedly anti-Christian 'Vulcanist' theories of the Scot James Hutton and the French *philosophe* Nicolas Desmarest. As a clergyman fascinated by natural history, William Richardson eagerly became involved. His summer residence near the Giant's Causeway provided profuse empirical knowledge of basalt and encouraged him to intervene in debates about its origins, debates which questioned biblical accounts of the earth's formation.

To the limited extent that historians of science have noticed Richardson at all, his opposition to Vulcanist theories has seen him portrayed as a reactionary dogmatist hidebound by theological and political conservatism. For Gordon Herries Davies, for instance, the 'real reason behind his opposition to these theories [Hutton's and Desmarest's] was not geological so much as theological'. Political conservatism and religious reaction are seen as inevitably linked. After the French Revolution, it was 'little wonder that a priest in the minority established church' should feel threatened by modern geology's perceived atheistical tendencies.[3] Davis A. Young sees theology and Richardson's Anglican ministry as influencing his opposition to Vulcanist theories, which questioned biblical accounts of Creation and the age of the earth.[4] Contemporary events provide a context. The guillotining of the pioneering French chemist Lavoisier and the assassination by the United Irishmen of a clerical geologist, William Hamilton, made 'establishment figures' naturally fear any system tainted with atheism and cling tenaciously to outmoded beliefs.[5]

Irish geology itself features much less in the historiography than its English and Scottish equivalents.[6] Richardson's presence is even more minimal, though several scholars do refer to his caustic views on Hutton, who believed (correctly) that basalt was igneous (i.e. volcanic) and whose followers opposed Abraham Gottlob Werner's 'Neptunian' view of its sedimentary origins.[7] In this heated debate Richardson has been seen as a zealous Neptunian or 'waterman'. For Patrick Wyse Jackson, he was the 'disciple' of the Irish Neptunist Richard Kirwan, whom he outdid in criticising Hutton and Desmarest.[8] Gordon Herries Davies agrees that Richardson and Kirwan both disliked igneous theories and that the former's discovery of fossil shells (*Cornua ammonis*) apparently embedded in basalt at Portrush made him a 'vociferous waterman' who 'clearly leant towards the Neptunian viewpoint'.[9]

The main problem with these interpretations is that they all, if in different ways, represent Richardson and his scientific milieu as basically static. Wyse Jackson reckons that, until Richardson himself slipped below the earth's surface in 1820, he continued 'to the last to hold onto his outdated geological opinions'. To be fair, what little has been written on Richardson usually refers to him in passing in discussions of other topics; this said, however, such second-hand commentaries give an inaccurate impression of the man. They are based on a limited range of sources which encourage interpretations of Richardson as intellectually moribund and his science as anachronistic. Wyse Jackson does note that Richardson produced a 'fairly constant stream of publications between 1802 and 1812'.[10] Yet even this does not do him justice. He continued geological publishing until December 1817 and was stopped only by illness. In 1818, despite being 'distressed by the stone', he was still thinking about stones, telling Greenough of 'A Philosophical Tour of the northern basaltic coast of Ireland' ready for publication by the Geological Society.[11] This chapter confronts the stasis issue head on by reconstructing Richardson's geological milieu from as complete a range of publications as possible and supplemented by unpublished manuscripts and private correspondence.

Such an exercise raises many questions. If we do not accept that Richardson was a curmudgeonly cleric trapped in outdated thinking, then what was he? Can his work be interpreted in the context of the development of geology as a discipline? An older historiography looked back from modern geology to a benighted past where men, warped by their subordination to prevailing religious beliefs or methodological naivety, reached what turned out to be the wrong conclusions. However, pioneering geologists are now examined in their contemporary context, facilitating understanding of their methods and the state of knowledge. Such research focuses more on their disagreements about evidence rather than who was eventually right or wrong.[12] Roy Porter sees Richardson's views on strata as indicative of contemporary thinking which reconfigured ideas on landforms, fossils and minerals. Porter also applauds his intellectual flexibility. Although opposed to Hutton, Richardson accepted that the earth had undergone massive denudation, unlike the 'scriptural geologists', who denied any 'naturalistic Enlightenment views of an active earth'.[13] This approach will be adopted here; however, much more needs to be done in interpreting Richardson's geology.

Three inter-related aspects stand out. Richardson can be read as a provincial scientific amateur fortuitously close to a major international geological phenomenon, the Giant's Causeway, trying to break into the charmed circle of metropolitan science. Secondly, again with a wider audience in mind, his geological publishing can be interpreted as an attempt to control the meaning of the knowledge that lay at his doorstep. This relates to the development of geology itself. Even if Richardson remained purely Neptunian, which he did not, Vulcanism's eventual triumph should not detract from the former theory's importance. When aboard the *Beagle* Charles Darwin had a geological book written by Werner's pupil Alexander von Humboldt.[14] Moreover, Richardson wrote during a pivotal time. Martin J. S. Rudwick has argued that the early nineteenth century was vitally important, as it foreshadowed a burst of 'exceptional fertility' from 1820, when the earth's immense antiquity was generally accepted. Finally, refusal to stereotype Richardson as an intellectual dinosaur raises broader questions about religion and science. Recent research rejects anachronistic assumptions of an automatic mutually exclusive divergence between science and religion.[15] If we unpick the symbiotic relationship between his religious outlook and scientific method, it becomes harder to sustain interpretations which ground Richardson's geological ideas simply in theological conservatism. As the real man of rock comes into view, the stereotypical reactionary Ulster cleric has as much substance as Fin McCool, the mythical giant who supposedly created the Causeway. Like the Neptunian theory of the earth, Richardson's geological activity had a beginning, a middle and an end.

Early geological publishing

Though Richardson's actual geological publishing began as the new century dawned, his interest in natural history ran deeper. The Giant's Causeway had long been a site for scientific visitors and it featured in early editions of the Royal Society's *Philosophical Transactions*.[16] Since the 1780s Richardson's summer sojourns at Portrush had made him intimately familiar with the area. His discovery of fossils in the late 1790s may have led to his opening a correspondence with other geologists. This has not survived, but its effects can be seen in visits by leading proponents of both the Neptunian and the Vulcanist theories. In 1801 the Swiss Neptunist Professor Pictet stayed with Richardson. Around the same time a Vulcanist, Professor Hope of the Royal Society of Edinburgh, also visited the Causeway. Richardson corresponded with Hope and sent him geological specimens and a memoir.[17] Around 1800 Richardson also sent specimens to a famous Neptunist, Sir Joseph Banks, President of the Royal Society. Banks asked if Richardson wanted to present a paper to the publication committee, but he hesitated, claiming that his observations were not repeated frequently enough 'to lay them before the public through so respectable a channel'.[18]

These tentative forays into metropolitan science suggest Richardson's yearning, as a provincial amateur with detailed knowledge of an international geological site, to cut a figure with the Edinburgh and London Royal Societies. But there were problems. Despite his knowledge and the Causeway's reputation, being remote from the centre meant that Richardson encountered the same proximity problem that bedevilled his political activities. Also, as his strategic hesitation before Banks shows, he wanted to ensure that the knowledge was correct or at least constructed in such a way as to be acceptable. Moreover, even assuming a fair wind, both Royal Societies had a considerable gestation period between acceptance of a paper and its appearance in their published transactions. Indeed, Richardson was not published by the London Royal Society until 1808 and the wait for the Edinburgh publication was almost as long. His memoir to Hope was not read until 1803 and did not appear in the Edinburgh Society's *Transactions* until 1805. Once embarked on a project, Richardson's temperament was such that he would try to short-circuit the process by trying other avenues to reach the centre.

The proximity problem was a matter of geography – Richardson was physically distant from metropolitan science – but it also concerned how the knowledge was transmitted. Visual media could enhance the printed word and, as the extensive circulation of engravings (1742–43) of Susanna

Drury's famous Causeway paintings showed, illustrations could help spread geological information.[19] Richardson understood this perfectly and had drawings and engravings made for several publications.[20] Though these illustrations were not printed, the intention was there, and several essays on basalt did appear in the prestigious *Nicholson's Journal* and in Tilloch's *Philosophical Magazine*, which was sold in London, Edinburgh, Glasgow and Dublin. In 1802 *Nicholson's Journal* carried a short essay on the 'basaltic coast of Ireland' addressed to Richardson's friend, George Knox MP (Thomas Knox's brother).[21] Both men had toured the Causeway in August 1800 examining whyn dykes, which Richardson believed continued under the sea to the Hebrides. Richardson's essay argued that these 'wonderful walls' were 'utterly irreconcilable to any of the theories invented by the cosmogonists'. Features like these which interrupted strata opened conjecture about the changes 'our globe has undergone since its creation'.[22] A second article, 'On basalts', appeared in the *Philosophical Magazine* in 1802. This was a reprint of a piece Richardson had written for the *Belfast News-Letter* ostensibly publicising a Causeway painting, but using the opportunity to discuss basalt. The visible evidence disproved various theories. Richardson's belief that the Causeway was part of the earth's original strata invalidated Desmarest's opinion that it was formed by a stream of lava running down from a volcano. The regularity of the basalt columns, which Desmarest said was caused by lava currents cooling, was unlike anything in genuine volcanic areas. Richardson also objected to the contention of the German geologist Erich Raspe that their shape was due to salt water acting on cooling lava, arguing that the evidence showed that most columns had no contact with the sea. Turning to Hutton's view that basalt was fused and consolidated lava forced upwards by subterranean fire, such an analysis would lead observers to expect broken strata, but the strata were disposed in a regular way. Kirwan's notion that the columns were formed by desiccation of rock was also incompatible with the facts.[23] These were polite interventions as Richardson waited for the English or Scottish Royal Societies to bestow recognition. This cautious provincial steered carefully. Aware of current controversies, he advanced only gentlemanly criticism and affected to appear uncommitted about theory. But beneath this disinterested facade he had the very thing he attributed to all theorists: a desire for fame.

Richardson's memoir to Hope, which was eventually read to Edinburgh's Royal Society in 1803, maintained the polite, non-committal tone of 1802. It detailed the discovery of 'siliceous basalt' at Portrush, which differed in its fine composition from the surrounding coarse basalt and contained fossils; Pictet believed it was a type of basalt with more silica than usual. Richardson knew that some mineralogists had denied it was basalt at all,[24] and did not venture an opinion, merely stating that he had encountered it only beside basalt. He adverted to the samples he had sent Hope, indicating the singularity of the whyn dykes and basalt with cavities containing fresh water. Earlier Causeway visitors, Dr Hamilton and John Whitehurst, had said they found basalt with 'bladder holes', thus suggesting it had volcanic origin but, apart from these specimens, Richardson found no cavities in Antrim basalt. While implicitly opposing the volcanic theory, again there was no overt criticism.[25]

However, alongside these British interventions, Richardson published memoirs with the Royal Irish Academy in 1803 which contained a marked difference in his treatment of theory and theorists. A 'letter' to Bishop Percy on whyn dykes in the Academy's 1803 *Transactions* noted that Scottish naturalists were interested in Hebridean dykes which, Richardson reiterated, originated on the Irish coast. For him, the dykes disproved the theory that basalt was volcanic. They were comprised of basalt, which, as the lack of volcanic material in Antrim proved, could therefore not have igneous origins. Adjacent dykes exhibited exterior and interior changes in composition, circumstances which were also incompatible with them having once been molten lava. However, Richardson progressively bore down harder against Hutton, borrowing a term from religious contention to describe the two 'sects', Vulcanists and Plutonists (who also believed that basalt was volcanic in origin), who, he argued, 'have of late taken possession of all the basalt in the world'. They were motivated more by egoism than empiricism, as 'a favourite theory is an adopted child, that must be maintained'.[26]

A second paper on this theme was read on 2 May 1803. Now Richardson lashed the Vulcanists with his pen with the same ferocity as he battered basalt facades with his geologist's hammer. This paper, on the inconsistencies of Hutton's theory, soon degenerated into a furious diatribe, associating Huttonians with atheism and revolution. Proud theorists could not accept that nature's secrets were concealed from them. And there was collateral damage. Richardson gave a contemporary twist to Bacon's dictum that 'every enquirer carries his own idols'. Theories were detrimental to scientific knowledge *per se*, but Bacon could not have predicted their impact in the hands of 'anti-Christian conspirators'. Citing the French royalist *émigré* the Abbé de Barruel, Richardson claimed that these 'false doctrines' would promote infidelity.[27]

Augustin de Barruel was a leading conspiracy theorist who had traced the French Revolution's origins to the Illuminati, the shadowy European offshoot of Freemasonry. Though today such interpretations seem bizarre, Barruel was taken seriously at the time by, among others, Edmund Burke and John Robeson, Professor of Natural History at Edinburgh and Secretary to its Royal Society.[28] Barruel's work was translated into English in 1797 by Robert Clifford and widely circulated. In 1798 Clifford had published his own pamphlet showing how Barruel's interpretation applied to Britain and to Ireland, where, it was argued, the United Irishmen paralleled the Illuminati. Barruel had claimed that the *philosophes* behind the *Encyclopédie*, a key Enlightenment text, were really revolutionary conspirators motivated by atheism to disprove biblical accounts of Creation. Warming to his familiar theme of misrepresentation, Richardson warned of dangerous doctrines disingenuously concealed 'under the mask of mere physical opinions'. Such devious conspirators would insinuate pernicious ideas into worthy occupations like natural philosophy. Though stopping short of accusing Hutton, Richardson claimed that his theory, as it became more widely known, was uncontrollable and dangerous in the wrong hands.[29]

Richardson ridiculed the Huttonian theory. He argued that basaltic strata predated any volcanic eruption and mocked the idea that the world's original surface was washed to the sea bed to harden into horizontal rock strata before being blown by explosions to the surface to form the mountains of 'a new world'. Hutton's dictum that the world had no beginning or end was linked to the Illuminatis' symbols of temporal eternity, the circular serpent and the phoenix. Hutton had died in 1797 but his followers were very much alive and included such heavyweights as Dr Hall and Professor John Playfair of Edinburgh, who had published *Illustrations of the Huttonian theory of the earth* in 1802. Always alert to chinks in his opponents' intellectual armour, Richardson highlighted Playfair's discomfort with atheistic interpretations of Hutton. Turning the screw, he implied that Hutton only followed the French *Encyclopédists*, Diderot, d'Alembert and Voltaire, who had questioned biblical accounts, and Desmarest, 'the father of the volcanic theory', who had even claimed that it 'may give the lie to Moses'.[30]

The rhetorical fireworks which opened this Irish publication stand in marked contrast to Richardson's measured British interventions as he awaited the support of the Royal Societies. This raises questions about his motives and the construction of geological knowledge. It could be argued that, as Richardson was known in Dublin circles from his Trinity days, he may have felt emboldened in attacking Hutton from this location and that such an assault would go down well with people, like himself, who had experienced actual conspiracy in 1798. Yet to concentrate on this startling opening, consisting of thirteen pages of polemical pyrotechnics, as several historians have done, distracts attention from the other forty-six pages, where Richardson's political and religious lacerations settled down to the kind of measured intellectual analysis which characterised his British publications. Lengthy technical sections described whyn dykes as empirical evidence refuting the Vulcanist opinion that these were lava veins. Hutton's notion of denudation was contrasted with the findings of Kirwan, who had collected rocks which showed no sign of diminution. The proposition that soils were composed of materials deriving from erosion of solid land was disproved by Kirwan's demonstration that they had a different composition from the underlying strata. Though Richardson hoped that specimens of the fossil-yielding 'siliceous basalt' would convince Playfair that it was once true basalt, he denied Playfair's allegation that his fossils had fuelled Neptunianism. He had 'never insinuated that … Neptunists had penetrated further into the secret of Nature than other Theorists' and could demonstrate neutrality in the theory wars by the samples Hope received, which 'seemed favourable to the Plutonic System'.[31]

Therefore Richardson's roaring anti-Vulcanist overture seems at odds with the laboured recitative in the body of the work, unless it is seen as a response to the rhythms of the scientific establishment. To interpret it thus allows us to place it in the context of his overall scientific ambitions. This explains an exchange of correspondence with Sir Joseph Banks around this time. In February 1803 he sent Banks a copy of the Irish whyn dyke memoir and offered specimens of basalt with freshwater cavities, similar to those he had sent to Hope. As we have seen, in Richardson's Scottish Royal Society memoir he adopted the persona of the innocent provincial empiricist whose value lay in his knowledge of and access to the Causeway. Now he calculated that playing one Royal Society off against the other was the best way to achieve recognition and pointed his evidence in a more theoretically partisan direction. The specimens represented 'an important Geological Fact' for Banks's 'Neptunian friends'. Richardson followed up by playing the national card. Banks heard how he was 'perpetually provoked and mortified' at 'puffs' about Hutton that the Scottish 'bandy backwards and forwards … as if they were exulting in a *Newton* or a *Copernicus*'.[32]

Thus, rather than being an instinctive anti-Vulcanist diatribe, Richardson's Irish memoirs represent the twists and turns of a provincial geologist seeking wider recognition. Living in the north of Ireland, it was hard to keep step with the metropolitan music. Banks, who had many friends among Scottish intellectuals, sent Richardson a letter in mid-1803 which arrived after the Royal Irish Academy memoir had gone to press. Chastened, Richardson admitted that if it had arrived earlier it 'would probably have stopped my proceedings against the Scotch theorists'. He tried to limit the damage, nervously forwarding Banks engravings of Fairhead, a prominent headland on the north Antrim coast, and claimed that the memoir arose solely from a desire to protect Kirwan (who had argued against Hutton) from 'incivility', which would 'excuse any sharpness I may have been guilty of towards his aggressors'. Banks's encouraging reply may have eased Richardson's anxiety: he thanked him for the engravings and the memoir, which he had read with 'pleasure as well as profit'. Without commenting on Richardson's invocation of political conspiracy or evocation of nationalistic chauvinism, Banks made his position clear:

> Geology is now making rapid strides but the Plutonians who seem resolved ... to wreck & twist everything till it fits their bed of Procrustes retard instead of advancing many parts of their discussion while the patient Neptunian who is content to admit the agency of fire in all places where the consequences of its actions are demonstrable has a far greater prospect of enlarging the science he cultivates.

Banks disavowed any discouragement and suggested further memoirs for himself and his friends. The Fairhead drawing, showing that vast basalt pillars were composed of smaller ones, was a new and interesting fact. Banks agreed that the observable changes in strata must have been wrought by causes no longer operative, and were 'of a nature wholly different from all things we are acquainted', making 'all reasoning from analogy ... useless'.[33] Reading Richardson's writings in the context of his ambitions, we can see that the risks he ran seemed justified. He had the encouragement of Britain's premier man of science and an explanation for irregularities in basaltic strata. But again geography got in the way of geology.

Unsure of the market for basaltic topics, Richardson needed Banks's influence with London booksellers. In June 1804 he sent a prospectus on the basalts of the north of Ireland. This probably formed the basis for a paper which appeared in the Royal Society's 1808 *Transactions* and as a separate pamphlet, but there was still a gap between aim and realisation. However, this chasm also represented differences in the state of knowledge in Ireland and Britain. Richardson understood this and offered the Dublin Society copies of his Causeway drawings. But he admitted to Banks that, at home, 'we neither, think, talk or read on such subjects' and was even prepared to abandon any general treatment of the subject to continue the 'skirmish' in 'our own Transactions with the *Volcanists* [*sic*] and M: *Desmarest*, their Father who is still alive to defend himself'.[34]

Richardson was being polite: he had no intention of taking a backward step. Now validated in his more decided Neptunism, he sent drawings to Lord Macartney, including one of the Portmoon façade, accompanied by a paper describing its sixteen strata. This was followed by an analytical section echoing Banks's view on geological change and agency. At Portmoon, 'nature seems disposed to reveal her secrets with more openness than any other place'. Yet 'every new fact' which was aligned against 'the ignorance and presumptuous vanity' of the 'adepts in the art of world-making' raised fresh and unanswerable questions. What alterations had the world undergone since it was formed and consolidated? What agency caused the mighty effects observable in nature, and had they ceased? Did they resemble known causes whose effects were traceable? Or were they entirely different, 'as they appear to be infinitely more powerful in their operation?'[35] Richardson was unable to answer these speculative questions, but recognising their significance, he asked Banks to forward the Fairhead engraving to Earl Macartney. However the politician in Richardson had not been completely subsumed by the scientist, as he saw how associational linkages could strengthen intellectual compatibility. So did Banks, who copied the Fairhead engravings and decided that he would nominate Macartney for membership of the Royal Society's Council.[36]

Now suitably emboldened by Bank's recognition, Richardson resumed his 'skirmish' with the Vulcanists in a series of papers to the Royal Irish Academy in December 1804.[37] Again he conflated atheism with Desmarest's theory that basalt was volcanic, citing the Abbé Barruel. The real danger came from the theory's wide influence. With Desmarest's seeming science exposed, the public should now reconsider the volcanic theory itself. Factual evidence showed that basalt had been found without volcanic scoria in places as far apart as Brazil and Staffa, as well as in Antrim. In Baconian reasoning, Desmarest's induction was flawed. Like Sir Richard Musgrave, he grounded his ideas on an assumption, in this case that basalt was igneous, and then sought supportive evidence to clinch the argument. Desmarest had never visited Antrim, but based his analysis on Susanna Drury's Causeway painting and mistook a conical hill for a volcano. Richardson cited empirical evidence against various Vulcanists' 'facts and opinions', gleefully pouncing upon inconsistencies in their arguments. Dr Perceval, Trinity College chemistry professor, had visited Richardson and observed that what the Vulcanist St Frond called lava currents were only ordinary ravines. Other absurdities were highlighted. John Whitehurst 'never got off his horse' when he visited the Causeway and what he supposed were volcanic cavities were actually manmade kilns for burning kelp and Richardson had a drawing captioned 'natives at the Giant's Causeway making lava'. Using formal logic, Richardson questioned the postulates on which St Frond and Dolomieu based their reasoning. Although these theorists were approved of by 'men decidedly attached to the religion they profess', they were fellow conspirators with Desmarest, Condorcet 'and other leaders of the anti-Christian, anti-Monarchical conspiracy', yet British naturalists were still being duped by them. Finally, evidence like the *Cornua ammonis* allowed negative comparisons between genuinely volcanic areas and Antrim. 'Moses's advocates', he said, could counter the volcanic theory's implications for the world's age by proving that the earth 'was formed in a much shorter time'. Nature playfully 'delights in diversity', confounding any all-embracing theory.[38] This relentless accumulation of facts obscures the fact that Richardson had no real argument. His aim, it must be remembered, was about lifting himself from provincial obscurity into the metropolitan limelight, a wish for recognition soon to be gratified.

National recognition

Banks may have turned a blind eye to Richardson's excesses but others in London did not. The June 1805 edition of the *Critical Review* contained acerbic commentaries on his Royal Irish Academy memoirs, claiming that Richardson's conspiratorial interpretation of Vulcanism was incompatible with 'philosophical freedom from prejudice'. The perceptive reviewer, striving to place Richardson intellectually, noted that he was definitely 'no Huttonian … but he certainly is an anti-Huttonian, and pursues that theory through every twig of its ramifications, with a hostility that never arose in the regions of science'. Richardson's prolix classical allusions provoked a disgusting image: 'he bespatters his opponents with opportune quotations to be found in the mouth of every school-boy'. His argument against the volcanic origin of basalt was set against Sir James Hall's experiments which showed that basalt could indeed be melted.[39] He was accused of attacking Hutton for positions he did not hold and of larding his own work with theoretical discussion to compensate for lack of real observation. Evocating 'the celebrated but doubtful antichristian conspiracy on the continent' was merely a device to damn Hutton's acolytes by association. Only 'an unusual muddiness of understanding' could confuse 'the atheistical assertion that the world has no beginning and no end; and the Huttonian conclusion that the world's natural appearances revealed no vestige of a beginning, and no prospect of an end'. Using de Barruel to compare Hutton with 'the open infidelity' of d'Alembert was a credulous view held 'with the partiality of a pedagogue'. Noting how Richardson's prose moved from savaging Hutton and Playfair towards a 'universal quietus', the reviewer claimed in a waspish anti-Irish statement that this would surprise English readers, from whom such a resolution, having 'the happy power to unite friends … and to soften the asperities of enemies', marked the end of political hostility. Affecting to take Richardson literally, the review closed with a hilarious conceit, imagining him and Playfair having a 'mineralogical dinner' at the Causeway, seated on broken basalt pillars around a table formed from whinstone with *Cornua ammonis* as salt-cellars.[40]

Never one to wilt under criticism, Richardson blasted back in a pamphlet admitting that he had attacked Hutton and Desmarest after discovering 'symptoms of infidelity'. Reflecting current orthodoxy, he reiterated that theories were harmless in themselves but dangerous from the inferences which could be drawn from them. The Irish example showed how perilous it was to ignore this. As someone driven from his house by the United Irishmen, he had 'unfortunately been witness to more conspiracies than one, the existence of which was by *some* [emphasis in original] most pertinaciously denied, until the pikes bristled on our streets'. Such asperity perhaps relates to the fact that, since his anti-Hutton memoir, his friend Kilwarden had been piked to death on a Dublin street. 'This Aristarchus', as Richardson dubbed his anonymous reviewer after the third-century Homeric scholar, was part of the same plot. Any distinction Huttonians made between their mentor's argument about the world's lack of a beginning or end and the 'favourite atheistical doctrine' of its eternity was like Jesuitical casuistry. Picking up on the anti-Irish tone, Richardson decided that the reviewer, as a Huttonian, must be Scottish and affected to wonder at how such an outwardly religious nation could countenance a theory which had not 'made its way across the Tweed'. 'The sober part of the Scotch (for we all have our Illuminati)' should prevent such 'modern philosophers from insidiously stealing in their innovating doctrines'.[41]

Again, far from being a raving sectary, Richardson knew exactly what he was doing. Hutton's theory was influential in Scotland but its adherents were outnumbered by Scottish Wernerians, like Robert Jameson, who would (in 1808) leave the (Huttonian) Royal Society of Edinburgh to form, along with Banks, the Wernerian Natural History Society.[42] Secure in his Causeway knowledge, confident of Banks's support and knowing Scottish opinion was divided over the rival theories, Richardson goaded and provoked in order to capitalise on the main chance.

This opportunism also had a bearing on the development of the discipline of geology, as the further Richardson was drawn into the basalt debate the more he had to justify his stance intellectually. This went rather against the grain for someone who enjoyed argument for its own sake. Earl Macartney once perceptively compared Richardson to a footballer who, 'never daring to hazard himself, by kicking the ball, watched for an opportunity of tripping up those who ventured themselves boldly'.[43] Yet, perhaps stimulated by Banks's forensic questioning, Richardson moved inexorably towards a view of his own. William Drummond, the Vulcanist author of a topographical poem on the Giant's Causeway, observed 'that the Doctor pretends to be an enemy to all theory, though it may be easy to discover from his conversation that he has formed one of his own'. Drummond paraphrased this 'theory' as 'a powerful cause acting from above', but Richardson, like Banks, could not explain it.[44] His private views became public on an auspicious occasion for any Irishman, geologist or otherwise. Saint Patrick's Day 1808 marked the long-awaited triumph of Neptune and nepotism. Richardson made his debut with London's Royal Society in a 'letter' to a Cornishman who also had intimate knowledge of a geologically important landscape, Humphry Davy.[45]

This was the tangible recognition that Richardson had been waiting on and it forced him to address an issue he had skirted around or obscured by rhetoric: the linkages between science and religion. Richardson and Davy were friends since the latter's visit to the Causeway in 1806 on a geological tour of Ireland. Davy was a rising scientific star. He was elected a Royal Society fellow in 1803, awarded the Copley medal in 1805 and made the Society's joint Secretary in 1807.[46] He was also a romantic poet, a friend of Wordsworth and Coleridge, with a profound interest in natural theology.

Natural theology has been defined as 'the practice of inferring the existence and wisdom of God from the order and beauty of the world'.[47] This doctrine had emerged from a university-based humanist tradition which encouraged study of the earth in various disciplinary contexts.[48] Natural theology's roots went back to Plato and it influenced key figures of the seventeenth-century scientific revolution, Sir Isaac Newton and the Irish scientist Robert Boyle, one of the founders of the Royal Society. It was held to complement rather than overturn revealed theology as the 'Book of Nature' was seen to supplement the 'Book of Scripture'. Natural theology also complemented the eighteenth-century spirit of Enlightenment and allowed for human reason: John Locke's *Essay* of 1689 had suggested that the idea of God was learned.[49] By the early eighteenth century, natural theology was considered orthodox in the Church of

England. It had produced John Ray's *The wisdom of God manifested in the works of Creation* (1691) and William Derham's *Physico-theology* (1713) and, more recently, William Paley's influential *Natural theology* (1802).[50] It was a tolerant doctrine which comfortably accommodated science and religion. Ray's work recognised apparent imperfections and failures of design in the world. Eighteenth-century science was 'saturated with natural theology', which offered a *via media* between doctrinaire zealots who associated it with deism and radical sceptics like David Hume and the *Encyclopédists*.[51] It was also a potentially unifying force, the appeal of which extended beyond liberal Anglicans and Unitarians to evangelicals like Thomas Chalmers, who equated empirical geology with theology, saying 'give us the facts' to understand the ways of God.[52] Awed by the sublime Causeway landscape, Davy reflected later, 'what is the end of our existence if it be not to investigate the wonders of creation, to understand the works of God?'[53]

On a scientific level, Davy's work has been described as a conservative reformulation of Enlightenment ideals, linking the maintenance of social order with the aesthetic appreciation of the natural world bestowed by science, without going to the radical lengths of Joseph Priestley, who linked scientific advance with political emancipation. Davy differed from Banks, who was bound to the ideas of Francis Bacon and utterly opposed to scientific theorising. Yet between them Davy and Banks epitomised enough continuity from eighteenth-century to early nineteenth-century science for the Reverend William Richardson of Clonfeacle to genuflect towards the Royal Society's Somerset House in London.[54]

The Royal Society therefore seemed to be the natural home for Richardson's ideas. Significantly, he opened his Royal Society 'letter' by defining his understanding of natural theology, based on John Ray's view of nature as 'the wisdom of God in the creation of the world'.[55] However, how could he reconcile this with the rugged and fantastic landscape which lay outside his door? Both Richardson and Davy believed in an ordered universe and struggled to explain natural 'inequalities' like the precipitous basaltic facades. How had they been formed? Banks had originally raised this problem with him in 1804. Some commentators complacently believed these beautiful features were simply the work of God in the Creation. But Richardson thought that interruptions and resumptions in the strata revealed that, though once continuous, they had been broken by some later convulsion. Consciously avoiding using the word God, he acknowledged disagreements about which direction the 'power' operated in, as well as about its means or agency. Some (like Hutton) were wrong to believe it acted upwards from the 'bowels of the earth'. If strata had been disturbed from below, they should appear in an oblique direction, yet basaltic strata lay parallel to each other. Richardson followed Paley in criticising the French natural philosopher Georges Buffon, who was also mistaken in his belief that natural forces continually destroyed and rebuilt the world's surface. Richardson could find no evidence of such action. He also countered Playfair's view that irregularities resulted from river erosion or some other general decomposition with evidence that rivers could not reach all areas. The basaltic district's surface had been altered from its original form by some immensely powerful agent and its interrupted strata were once continuous. These facts led Richardson to conclude that the interrupted strata were produced by an 'unknown cause', acting from above, removing material where it touched, but leaving other features undisturbed. The basaltic hummocks that Davy and Richardson observed in 1806 were the remains of the original surface, untouched by the 'powerful agent'.[56]

This important letter reveals aspects of Richardson's work with wider significance. No matter how he explained it, by accepting that the earth's surface continued to change after its original formation, Richardson contributed his mite to what would become one of the defining issues in the development of geology – the debate about the age of the earth. In the context of the Vulcanist debate, Richardson expected a riposte from Playfair, but none materialised. There was a profound irony here: Richardson's wider currency rested on his first-hand knowledge of the geological singularity of the Antrim coastline; yet, at the same time, geographical isolation cut him off. He asked Greenough, 'Has he [Playfair] fallen upon me as threatened?' but attributed this need to ask to his own distance from the centre of debate: 'my friends must tell me for in my retirement I never hear of such things'.[57] What Richardson saw as a disadvantage Davy would have seen as advantageous, claiming in a lecture that 'it is from minds nourishing their strength in solitude, and exerting their strength in society, that the most important truths have proceeded'.[58] Isolated in Antrim, Richardson's views were changing. Though his Royal Society letter tactfully avoided criticising the Neptunian theory, possibly in deference to Banks, he told Greenough privately of his determination to fight the Neptunians, 'as I long to do it'. Richardson was gathering facts on strata for the Belfast Literary Society, which would allow him to 'trim Jameson whom I hold much cheaper than you seem to do', by showing that 'the world was never made by Werner of whom he displays such a silly reverence'.[59] This eventual rejection of the Neptunian theory tells us much about the direction of Richardson's thinking and the impact of natural theology. For man to find nature's secrets, it was necessary to go to the natural world for the facts, not, as with the Vulcanists and the Neptunists, to concoct a theory from their own minds. This supposed self-referential tendency was what Richardson had in mind when he quoted Bacon that 'every enquirer carries his own idols'.[60] From now on, the derisive term 'world-makers' was applied indiscriminately to all theorists.

As he had hoped, the Royal Society memoir attracted notice and gave pretexts for further publishing. John Farey noted in his survey of Derbyshire, commissioned by the Board of Agriculture, that his views on the arrangement of strata coincided with Richardson's.[61] In a letter to Tilloch's *Philosophical Journal* Richardson responded to Farey's flattering notice and elaborated his view of the unknown agency's power to produce 'diminutive works of nature', from basaltic facades to mountains. 'Cosmogonists' seeking to understand the earth's original formation missed the importance of how strata were subsequently arranged. Emphasising that 'I limit myself to facts', his exploration of Knocklaid mountain had convinced him that it, and by analogy all mountains, were not formed by the Vulcanists' 'stupendous constructions of mighty agents' but were 'the scattered remnants of a diminished world' left after the unnamed force carried the rest away.[62] While he wrestled with the high intellectual problems raised by the evidence, Richardson also addressed an issue at the opposite end of the spectrum, the expansion of the audience for science.

Historians have long recognised the role of popularisers like Humphry Davy, whose Royal Institution lectures made science a compelling spectacle.[63] Scientific tourism can be seen in the same light. The Giant's Causeway had long attracted elite scientific pilgrims, but by the early nineteenth century it hosted a wider audience of the curious, as eager to be thrilled by the rugged wave-lashed environment as they were for geological enlightenment. Unlike the solitary scholars of a previous generation, these visitors came in groups and expected an exciting guided tour. Richardson understood this early manifestation of cultural tourism and 'adopted' one of these guides, Alick Macmullan, a highly intelligent but uneducated local. He provided instructions to ensure that visitors saw geological features like the whyn dykes, as well as attention-grabbing curiosities such as the 'Giant's Chair'. Alongside geological explanation, and adding to the drama, their romantic sensibilities were to be engaged by hearing that no man had yet discovered the depth of the dykes. Macmullan was told to give short shrift to the usual theories. 'Neptune, Vulcan and Pluto, have each put in their claims, which are supported by their respective friends', but he was to act the stage Irishman, defying his gaping audience 'to prove … that it was not made by Fin McCool'. For his many Scottish visitors, Macmullan would imply Huttonian absurdities by archly whispering in their ears that rumour had it that 'it is all lava … melted at the bottom of the sea!' After this bravura performance, on receiving the 'accustomed gratuity', Macmullan should turn the coins in his hand, 'look grave, scratch your head and observe that it might be quite sufficient for any of the impertinent lazy fellows now crowding around them, but for a man of literature, it is rather little'. Visitors wanting information on shells embedded in basalt were to be sent to Richardson at Portrush.[64] Two points emerge from this. Firstly, the jocularity concealed a consistent purpose. Even if Richardson archly enlisted Fin McCool, he genuinely wanted to disabuse visitors that any theories could explain what they had seen, and to promote his own notion of some primeval force. Secondly, John Ray's natural theology had invoked nature as God's agent.[65] Richardson developed the idea though refused to speculate on the origin of the force, instead using terms which emphasised its power, which chimed with Burke's conception of 'sublime' features in nature, the magnitude and vastness of which produced astonishment and awe in the beholder.[66]

These themes were continued in a statistical survey of county Antrim commissioned by the Dublin Society. This was the 'scattered remnants' of a more ambitious project by the Belfast Literary Society, which had been aborted due to the inertia of some of its contributors.[67] Richardson's input to the statistical survey was an expanded version of an unpublished 1811 memoir to the Geological Society[68] with an additional 'letter' on zeolite and ochre. He argued that neither the reddish ochre found in basalt nor the presence of zeolite (a mineral now known to form in cavities in lava flows) suggested that it was once fluid lava, due to zeolite's properties of containing water and being fusible in moderate heat. Richardson noted how knowledge of strata had practical benefits and could help locate sites for quarrying and mining. He also repeated his 'downward force' analysis, included a vividly descriptive travellers' itinerary detailing whyn dykes and the arrangement of strata, and delivered an onslaught on 'the Volcanic, the Huttonian and the Neptunian' theories.[69]

The Royal Society letter undoubtedly marked the high-point of Richardson's geological publishing. Important as this was in terms of a provincial getting national recognition, the underlying process of engagement with geological debates has wider significance regarding where Richardson's geology fits in with the development of the discipline. His attention-grabbing anti-conspiracy rhetoric and evocation of national sensitivities could also be seen as compensating for his geographical isolation: the further from the centre, the more noise was needed to get noticed. However, personal ambition aside, being drawn into scientific debate forced Richardson to consider the implications of his empirical methodology and how they squared with natural theology, which eventually led him to defend his position and criticise all theories. This intellectual endeavour contrasted with his opening salvoes, where he had used geological debates instrumentally to further his own ambition. The close of Richardson's geological work may be dated from 1811 to 1818 and, unlike his earlier efforts, was characterised by local publishing and private correspondence. His interest in agricultural botany was becoming predominant, but the narrowed scope of his late geological work helps draw the strands of his thinking together.

'Downward force' – late geological work

Richardson's letters to Greenough in this final stage of his geological activity were characterised by the same championing of empiricism over theory. Greenough had heard that the Mourne Mountains would make a good case study. Wernerians thought they could find support for their theory there, while Vulcanists thought this mountain range proved Hutton; but Richardson thought both were 'equally ignorant in world-making'.[70] An article in the inaugural edition of the *Newry Magazine* carried out his threat to lash all theorists. Evidence from Portrush invalidated every theory. Each side of its peninsula differed in composition and abounded in marine fossils, which disproved Vulcanism, and, though fossil evidence ostensibly supported the 'Neptunian cosmogonists', even Kirwan could not explain strata formation empirically. The universality implied by the Neptunian claim that strata formed by precipitation at 'the bottom of a widely extended fluid' did not tally with the diversity of this small area. Conversely, the perpendicular basalt columns on one side of the peninsula and the uniform inclination of strata on the other side disproved volcanic origin. Thus the peninsula could not have been formed by violent subaqueous convulsions to 'form a new world out of an old', according to the Huttonian or Plutonic theory, still 'so pertinaciously supported by our *Caledonian* brethren'.[71]

Richardson turned briefly from his growing agricultural interests when the Countess of Gosford asked for his comments on the translation by the Scottish Wernerian Robert Jameson of Georges Cuvier's *Theory of the earth*. The French scientist posited a catastrophist view on geology but Richardson commented '*sharply* holding Mr Cuvier and the Neptunians very cheap *cosmogonists*'. Two 'letters' to Lady Gosford in the

Newry Magazine contained 'a development in my old ideas on the cause of the inequalities of our surface'.[72] They confirm that Richardson now fully concentrated his ire on the 'Neptunians, or as they are now sometimes called, the Wernerians'. Cuvier's work was dismissed as 'a mass of geological facts … arbitrarily produced, without any reference to the places they are generally found' and were 'mere inventions to countenance the Neptunian theory'. Like all theorists, the Wernerian 'world-makers' 'looked inwards' to their own imaginations, not to 'the face of Nature'. Factual evidence of diversification undermined their assumption that general and uniform causes had formed and arranged the world.[73] The second letter attacked the Neptunian theory's premises, which required 'the aid of revolutions and convulsions'. Instead of resulting from catastrophic upheavals, Antrim's strata showed signs of only minimal disturbance, in a slight tilt from their original horizontal position. Kirwan's view that strata formed by precipitation from a 'chaotic fluid', initially powerful but subsequently abating, leading to gradual changes in their composition, was 'nothing like Nature'. Individual strata were found to be uniform. Knowing that Irish landowners, like Lord Gosford, were interested in the mining potential of their estates, Richardson warned them to shun 'mountebanks', 'coal-finders' or speculative theorists, but instead to take advice from 'plain honest men'. Once accused of being 'contentious and quarrelsome' against 'vain theorists', though he had given up explaining early processes of the earth's formation, he could now defend his intellectual credibility with his idea of a downward force. Though some believed this to be wilder than the theories he scorned, Richardson insisted that empirical observation made the distinction between original and subsequent operations abundantly clear.[74] Further *Newry Magazine* essays hammered home the point by comparing Antrim and Saint Helena. A traveller's account stated that the island was formed by 'protrusion from the sea bottom', but Richardson twisted the evidence to argue that the mountains of Saint Helena resulted from the 'removal of contiguous materials', thus supporting his 'diminished world' analysis.[75]

Signs of the old ambition were still there. Copies of the essays were duly sent to Greenough, as President of the Geological Society, and to John Kidd. He was Professor of Chemistry at Oxford and supported Paley's natural theology. Richardson felt Kidd's views on mineral veins to be compatible with his own on whyn dykes.[76] Greenough had challenged Richardson's claim that the denuding agent acted from above, but Richardson defended himself, saying that the location of strata at the foot of facades and dykes suggested that they had never been disturbed, whereas they would have been if the agent had acted horizontally.[77] Richardson still denied developing a theory, protesting that his conclusions, however wild, were not his responsibility. He was a merely a 'passive agent': 'Nature … gave the premises and science [and] taught us to draw conclusions'. This argument drew on natural theology and Richardson pursued it further by adverting to other barren extremities. The 'Kobi' (i.e. Gobi) desert was 'an untouched remnant of the original block of the world … left for the mighty agent … to perform his operations upon'. Physico-theologists who extolled 'the wisdom and goodness of God in the arrangement of our natural world' emphasised the benefits derived from natural inequalities, like verdant valleys and pure mountain air. Therefore it was not unreasonable to suppose that, in the fullness of time, 'the Almighty Disposer of all things' would 'call again into action the same agent', whose footsteps Richardson traced in Antrim, to 'improve' the desert.[78]

Richardson's final *Newry Magazine* contribution came in 1817, in a letter complaining about Davy's successor as Royal Institution lecturer, W.T. Brande. He had conjured one of Richardson's old Vulcanist demons, by publishing lectures eulogising John Whitehurst, whose 1778 *Enquiry into the original state and formation of the earth* claimed basalt's igneous origin. In a familiar tactic, Richardson humoured his opponent to undermine him. A footnote recalled how, when Whitehurst had originally published, his supporters included Frederick Hervey, the eccentric Earl Bishop of Derry, who had met Desmarest. A droll anecdote related how Hervey was once accosted by a beggar demanding he help a poor 'cratur' and apparently replied: 'I have found the lava, but show me the crater'. Richardson said he had written a memoir against Whitehurst around 1798 but suppressed it, believing the volcanic theory was becoming unfashionable. However, Brande's praise of Whitehurst, and his influential position, spurred Richardson into renewing his quarrel 'with the world-makers'. Also in 1817, a pamphlet was published in Newry, with copies for the Geological Society, based on his tour with Davy of the basaltic coast of Derry and Antrim, which reads as a mixture of scientific disquisition and tour guide.[79] Technical details were enlivened by descriptions reflecting the romantic influence. Richardson said that a partial peep at 'stupendous facades' at Bengore and Fairhead was available from the land; however, this 'grand display of superb scenery' was best seen by boat. As Carrick-a-Rede was 'a scene of singular beauty', where a rope bridge for salmon fishermen straddled the chasm between a precipitous shoreline and a rocky outcrop (as it still does now), Richardson was 'more terrified than amused by seeing from the boat the fishermen passing the swinging bridge, at such an enormous height above me'. A chasm nearby, dubbed 'Grace Staple's cave' after a young lady of a landed family who bravely accompanied him on a previous boat trip, revealed an inverted causeway lying beneath the 'deep sea green water'. They sailed further into the cave, where 'the scenery about us, especially when the gloom commences, is most beautiful'.[80]

Richardson's geological swansong in December 1817 was spurred by a memoir on north-eastern Ireland for the Geological Society by Dr Berger and the Reverend William Conybeare. Richardson hated their Wernerian tendencies but, as they had Greenough's support, trod warily. 'What will I do with that Conybeare of yours, so … good humoured, but shy of fighting?' However, he could not ignore fundamental disagreements. 'He, like other Wernerians, makes our world a whole … I say Ireland is a piece of mosaic, the patches are connected with one another'. Richardson was mortified by their 'very cheap' treatment of 'my country' and accused them of having 'borrowed from me, and with but little acknowledgement'. However, Richardson implicitly conceded his marginality. His riposte was 'likely to prove an abortion' and required toning down to treat the pair 'with more civility than they as geologists (I cannot call them philosophers) deserve'.[81] He contented himself with opposing Conybeare's contention that the world's present surface was caused by water erosion, which would occur horizontally, arguing that whyn dykes, stratified hummocks and vertical facades proved the vertical operation of the 'denuding agent', facts which had 'escaped the notice of the geological tourists'.[82]

Richardson's changing views

Thus William Richardson's geological activities opened with deference towards scientific visitors to the Causeway like Pictet and ended with his scornful dismissal of Berger and Conybeare as 'tourists'. It would be easy to dismiss him as inconsistent or reactionary. Yet Richardson's own geology, like his view of the world's surface, changed over time. It is therefore quite incorrect to categorise him from his early flaying of the Huttonians. Setting his public utterances against the intellectual and institutional context revealed in private correspondence allows a more nuanced interpretation, that these rhetorical outbursts were strategic. The cultural and intellectual necessity to grace the wider scientific stage was the mother of polemic invention. If these utterances are seen as rhetorical positioning rather than the statement of a scientific position, we can see just how knowledgeable this Irish provincial amateur geologist was about developments at the centre of the British scientific world. Richardson wrote when a clearer disciplinary estimation of the earth's age was emerging. His polemical stances eventually forced him to take a position himself, which has broader relevance in this formative period. As he told Macartney, finding facts to throw at 'adepts in the art of world-making' raised fresh and unanswerable questions.[83] These new ideas had a tortuous birth: Richardson's views yawed to and fro as dizzily as the rope bridge at Carrick-a-Rede. Beginning with tactful neutrality, then favouring Neptunism over Vulcanism, then criticising both, and eventually directing his fire against the 'watermen', though he turned out to be wrong, such a progression showed considerable intellectual flexibility for a man in his seventies.

These changing views forced Richardson to defend his scientific stance intellectually and to consider how it sat with his religious position. This brought two different modes of thought into contact. Firstly, the logical outcome of applying his store of empirical facts to debate pushed him in a broadly progressive direction, eventually to reject all catastrophist interpretations of the world's surface. Though he included himself with the 'advocates' of Moses, he was no scriptural geologist wedded to biblical accounts of Creation. The interaction of religion and science in his geology is much more complex than will allow simply bracketing him off as a theological and political conservative. His thinking was certainly influenced by natural theology, as indicated by his agreement with the physico-theologists on wastes awaiting the beneficent hand of God as divine improver and his conception of mankind's intellectual humility before the secrets of nature. While rational enquiry was man's proper response to the natural world, arrogant theorising was not. With the innocence of a child, nature playfully 'delights in diversity' yet could act with unimaginable force.[84] The God of nature and not the self-made god of human vanity was the key to understanding the earth. This fitted well into the provincial scientific context. Natural theology was influential with the liberal Presbyterians Richardson associated with in the Belfast Literary Society as well as that denomination's conservative and evangelical strands.[85] In struggling to describe the power and mystery of the 'downward force' and in his topographical descriptions there are signs of a romantic sensibility trying to free itself from the chrysalis of the eighteenth-century rationalist. Yet, despite connections with the Geological Society, beginning his publishing in London and ending it in Newry suggest that Richardson the geologist was increasingly ignored. The same cannot be said for Richardson the agricultural improver.

Notes

1 W. Richardson, 'On the strata of mountains', *Philosophical Magazine*, vol. 37 (1811), pp. 367–9.
2 Johann Wolfgang von Goethe, *Faust: a tragedy – parts one and two*, translated by R. D. MacDonald (London, 1988), p. 184; F. Bacon, *Instauratio magna: praefatico*, cited in M. Purver, *The Royal Society, concept and creation* (London, 1967), p. 144.
3 G. L. Herries Davies, 'Geology in Ireland before 1812: a bibliographical outline', *Western Naturalist*, vol. 11 (1978), p. 91.
4 Davis A. Young, *Mind over magma: the story of igneous petrology* (Princeton, 2003), pp. 35–6.
5 G. L. Herries Davies, 'Astronomy, geology and mineralogy', in T. O'Raifeartaigh (ed.), *The Royal Irish Academy: a bicentennial history* (Dublin, 1985), pp. 256–7.
6 Herries Davies, 'Geology in Ireland', pp. 79–99.
7 P. Wyse Jackson, 'Tumultuous times', p. 40; G. L. Herries Davies, 'The history of the earth sciences', in D. G. Smith (ed.), *Cambridge encyclopaedia of earth sciences* (Cambridge, 1981), pp. 12–17; Herries Davies, 'Astronomy, geology and mineralogy', pp. 252–4; Herries-Davies, 'Geology in Ireland', p. 88.
8 Wyse Jackson, 'Tumultuous times', pp. 43–4.
9 Herries Davies, 'Astronomy, geology and mineralogy', pp. 255–7; Herries Davies, 'Geology in Ireland', pp. 90–1.
10 Wyse Jackson, 'Tumultuous times', pp. 43–4.
11 Richardson to Greenough, 25 September 1818, UCL, Additional Greenough papers, 1506.
12 R. Rappaport, 'The earth sciences', in R. Porter (ed.), *Cambridge history of science, vol. 4: eighteenth-century science* (Cambridge, 2003), p. 435.
13 R. Porter, *The making of geology: earth science in Britain, 1660–1815* (Cambridge, 1977), pp. 165, 176, 204, 233.
14 Sandra Herbert, *Charles Darwin, geologist* (Ithaca and London, 2005), p. 58.
15 Martin J. S. Rudwick, *Worlds before Adam: the reconstruction of geology in the age of reform* (Chicago and London, 2008), pp. 2–3, 5; Thomas Dixon, 'Introduction', in T. Dixon, G. Cantor and S. Pumfrey (eds), *Science and religion: new historical perspectives* (Cambridge, 2010), pp. 1–2.
16 Porter, *Making of geology*, pp. 38, 40.
17 Marc-Auguste Pictet, *Voyage de trois mois, en Angleterre, en Écosse, et en Irelande* (Geneva, 1802), pp. 120–5; Richardson to Sir Joseph Banks, 8 February 1803, in N. Chambers (ed.), *The scientific correspondence of Sir Joseph Banks* (London, 2007), vol. 5, pp. 273–4; W. Richardson, 'Remarks on the basaltic coast of Antrim', *Transactions of the Royal Society of Edinburgh*, vol. 5, part 3 (1805), pp. 15–20.
18 William Richardson to Sir Joseph Banks, 8 February 1803, *Scientific correspondence of Sir Joseph Banks*, vol. 5, pp. 273–4.
19 M. Anglesey, 'The art of nature illustration', in J. W. Foster (ed.), *Nature in Ireland: a scientific and cultural history* (Montreal and Kingston, 1997), p. 501.
20 Andrew Caldwell to the Bishop of Dromore, 29 December 1801, in J. B. Nichols (ed.), *Illustrations of the literary remains of the eighteenth century* (London, 1858), vol. 8, p. 38.
21 W. Richardson, 'Observations on the basaltic coast of Ireland in a letter transmitted

from William Richardson DD by the Hon. George Knox MP', *Journal of Natural Philosophy, Chemistry and the Arts*, vol. 5 (1802), pp. 321–5; Richardson to Sir Joseph Banks, 29 January 1804, *Scientific correspondence of Sir Joseph Banks*, vol. 5, pp. 334–5.

22 Johnson defined the cosmogonist as 'he who describes the creation of the world'. See A. Chambers (ed.), *Samuel Johnson's Dictionary of the English language* (London, 1994), p. 155; Richardson, 'Observations on the basaltic coast', pp. 321–2, 325.

23 'W. R.', 'On basalts', *Philosophical Magazine*, vol. 13 (1802), pp. 129–36.

24 Herries Davies, 'Astronomy, geology and mineralogy', p. 255.

25 Richardson, 'Remarks on the basaltic coast of Antrim'.

26 W. Richardson, 'An account of the whynn dykes in the neighbourhood of the Giant's Causeway, Ballycastle and Belfast in a letter to the Bishop of Dromore', *Transactions of the Royal Irish Academy*, vol. 9 (1803), pp. 21–43.

27 W. Richardson, 'Inquiry into the consistency of Dr Hutton's theory of the earth with the arrangement of the strata and other phenomena on the basaltic coast of Antrim', *Transactions of the Royal Irish Academy*, vol. 9 (1803), p. 430.

28 Mirala, *Freemasonry in Ulster*, p. 18.

29 Richardson, 'Inquiry into the consistency of Dr. Hutton's theory', pp. 430–2.

30 Rudwick, *Worlds before Adam*, pp. 90–1; Richardson, 'Inquiry into the consistency of Dr. Hutton's theory', pp. 434–9, 452, 456, 475.

31 Richardson, 'Inquiry into the consistency of Dr. Hutton's theory', pp. 445, 447–8, 482–3.

32 Richardson to Banks, 8 February 1803, *Scientific correspondence of Sir Joseph Banks*, vol. 5, pp. 273–4.

33 Richardson to Banks, 29 January 1804, Banks to Richardson, 20 March 1804, *Scientific correspondence of Sir Joseph Banks*, vol. 5, pp. 334–5, 346–7.

34 Proceedings of the Dublin Society, 3 May 1804, Royal Dublin Society Library; W. Richardson, 'Letter on the alterations that have taken place in the structure of rocks of the basaltic country in … the counties of Derry and Antrim read before the ROYAL SOCIETY, March 17 1808', *Philosophical Transactions of the Royal Society* (1808); Richardson to Banks, 3 June 1804, *Scientific correspondence of Sir Joseph Banks*, vol. 5, p. 353.

35 Richardson to Macartney, 3 August 1804, PRONI, Macartney papers, D572/14/2a/26.

36 Richardson to Macartney, 29 October 1804, Banks to Macartney, 22 November 1804, PRONI, Macartney papers, D572/14/19, 30.

37 W. Richardson, 'On the volcanic theory part 1 Examination of Mr. Desmarest's memoir, in Transactions of the Academy of Sciences (1771); On the volcanic theory part 2 Examination of the facts and opinions, given by different advocates for the volcanic origin of basalt, who followed Mr. Desmarest: to wit, Mr. Faujas de St. Frond, Mr. Dolomieu, Mr. Whitehurst, Bishop Troll, Abbe Spalanzani, and Dr. Hamilton. Part 3 Arguments against the volcanic origin of basalt, derived from its arrangement in the county of Antrim, and from other facts observed in that county, communicated by the Rt. Reverend, the Lord Bishop of Dromore", *Transactions of the Royal Irish Academy*, vol. 10 (1806), pp. 35–108.

38 Richardson, 'On the volcanic theory', pp. 35–9, 41, 45–58, 65–76, 82, 91, 107.

39 *Critical Review or Annals of Literature*, vol. 5, no. 2 (June 1805), pp. 175–80.

40 *Critical Review or Annals of Literature*, vol. 5, no. 2 (June 1805), pp. 181–9.

41 W. Richardson, 'Review of two memoirs', *Transactions of the Royal Irish Academy* (1805), pp. 4, 6–7, 10–12.
42 Porter, *Making of geology*, pp. 149–55.
43 W. Richardson, *Second letter to the Countess of Gosford* (Newry, 1816), p. 6.
44 William Drummond to Robert Anderson, 7 June 1808, National Library of Scotland, Advocates' papers, MS 22.4.16, fols 79–80.
45 W. Richardson, *Letter on alterations that have taken place in the structure of rocks on the surface of the basaltic country in the counties of Derry and Antrim* (London, 1808).
46 Sir Hamilton Hartley, *Humphry Davy* (Menston, 1971), pp. 48–9.
47 M. D. Eddy and D. Knight (eds), 'Introduction' to W. Paley, *Natural theology or evidence of the existence and attributes of the deity, collected from the appearances of nature* (Oxford, 2006), p. ix.
48 Porter, *Making of geology*, p. 23.
49 Eddy and Knight, 'Introduction', pp. x–xii.
50 D. W. Bebbington, 'Science and natural theology in Britain from Wesley to Orr', in D. N. Livingstone, D. G. Hart and Mark A. Noll (eds), *Evangelicals and science in historical perspective* (Oxford, 1999), p. 127; James E. Crimmins, 'Paley, William (1743–1805)', *Oxford dictionary of national biography* (Oxford, 2004), online edition, January 2008, www.oxforddnb.com/view/article/21155, accessed 19 August 2011.
51 Eddy and Knight, 'Introduction', pp. xi–xiii.
52 Bebbington, 'Science and natural theology', pp. 120, 125–6.
53 Royal Institution, Humphry Davy's journals, HD/15/J, fol. 161.
54 J. Golinski, *Science as public culture: chemistry and Enlightenment in Britain, 1760–1820* (Cambridge, 1992), pp. 196–7; D. Knight, *Humphry Davy: science and power* (Cambridge, 1992), p. 37.
55 Richardson, 'Letter on the basaltic surface of the counties of Derry and Antrim', pp. 3–4.
56 Eddy and Knight, 'Introduction', p. xxvi; Richardson, 'Letter on the basaltic surface of the counties of Derry and Antrim', pp. 24–7, 33, 37.
57 Richardson to Greenough, 5 April 1809, UCL, Additional Greenough papers, 1423.
58 Cited in Golinski, *Science as public culture*, p. 188.
59 Richardson to Greenough, 18 May 1809, UCL, Additional Greenough papers, 1424.
60 Richardson, 'Inquiry into the consistency of Dr. Hutton's theory', p. 2.
61 Porter, *Makers of geology*, pp. 96, 136–7.
62 Richardson, 'On the strata of mountains', pp. 367–9.
63 R. Holmes, *The age of wonder* (London, 2008), p. 304.
64 W. Richardson, *Letter to the curious traveller, who shall visit the Giant's Causeway* (Coleraine, 1811).
65 Eddy and Knight, 'Introduction', p. xi.
66 Cited in C. Lavin and I. Donnachie, *From Enlightenment to Romanticism* (Manchester, 2004), pp. 9–12.
67 Richardson to Greenough, 18 October 1811, UCL, Additional Greenough papers, 1468; Dr Bruce's Address to the Belfast Literary Society, c. 1829, NLI, Bruce MSS, 20898; W. Richardson, *A description of the basaltic coast from Magilligan to Glenarm* (Newry, 1817), pp. 7–8.
68 Richardson to Greenough, 8 July 1811, UCL, Additional Greenough papers, 1464.

69 Dubordieu (ed.), *Statistical survey*, pp. 3–15, 24–6, 29–31, 37–51, 62.
70 Richardson to Greenough, 25 August 1811, UCL, Additional Greenough papers, 1467.
71 W. Richardson, 'Account of the peninsula of Portrush', *Newry Magazine or Literary and Political Register*, vol. 1 (1815), pp. 250–5; Richardson to Greenough, 25 October 1815, UCL, Additional Greenough papers, 1492.
72 Richardson to Greenough, 30 January 1815, UCL, Additional Greenough papers, 1488.
73 W. Richardson, 'Letter to the Countess of Gosford', *Newry Magazine*, vol. 2, no. 6 (January–February 1816), pp. 37–41.
74 W. Richardson, 'Second letter to the Countess of Gosford', *Newry Magazine*, vol. 2, no. 7 (March–April 1816), pp. 95–9.
75 W. Richardson, 'Saint Helena', *Newry Magazine*, vol. 2, no. 8 (May–June 1816), pp. 211–18.
76 Richardson to Greenough, 20 March, 14 April, 26 May 1817, UCL, Additional Greenough papers, 1495, 1496, 1497; Porter, *Making of geology*, pp. 1, 183; Eddy and Knight, 'Introduction', p. xxvi.
77 Richardson to Greenough, 20 March 1817, UCL, Additional Greenough papers, 1495.
78 W. Richardson, 'Saint Helena', *Newry Magazine*, vol. 2, no. 9 (July–August 1816), pp. 302–7.
79 'Dr Richardson's dissertation on Whitehurst's enquiry', *Newry Magazine*, vol. 3, no. 16 (September–October 1817), pp. 353–8; Richardson, *A description of the basaltic coast from Magilligan to Glenarm* (Newry, 1817); Richardson to Greenough, 30 October 1817, UCL, Additional Greenough papers, 1500.
80 Richardson, *A description of the basaltic coast*, pp. 2–5, 8.
81 Richardson to Greenough, 14 May, 2 August 1817, UCL, Additional Greenough papers, 1494, 1498.
82 W. Richardson, *Observations on memoirs published by the Geological Society respecting the north-eastern counties of Ireland* (Newry, 1817), pp. 3–7.
83 Richardson to Macartney, 3 August 1804, PRONI, Macartney papers, D572/14/2a/26.
84 Richardson, 'On the volcanic theory', pp. 58, 65–7, 74, 76, 82, 91, 107.
85 Andrew R. Holmes, 'Presbyterians and science in the north of Ireland before 1874', *BJHS*, vol. 41, no. 4 (December 2008), pp. 545–6.

4

The man of grass: agricultural improvement and public opinion

Some say (but impudence with some is law)
That Patriots now-a-days are *men of straw*;
Who dares the assertion is an errant ass:
For thou, *great Patriot*, art a *man of grass*.
('Ode to Richardson', 1815)[1]

In August 1818 a review in *Farmer's Magazine* ridiculed Richardson's *Essay on agriculture* (1818), in which, as in many publications, he claimed that Irish fiorin grass would yield huge hay crops in wintertime, provide nutritious food for livestock and convert bogs into profitable meadow. The reviewer's scorn focused on a memoir appended to the essay recommending fiorin to Archduke John of Austria, 'one of the most promising performances that ever issued from the press'. Recalling the Napoleonic wars, when France ruled Austria, the reviewer speculated whether fiorin might 'become a dangerous instrument … if His Imperial Majesty should ever again become our enemy'. Richardson's essay on a grass, which had aroused massive controversy, was unparalleled, 'except in the annals of empiricism – or lunacy'.[2] Accusations of insanity were not unknown for eighteenth-century improvers.[3] Richardson's contemporary, the Scottish natural historian John Walker, was dubbed 'the mad minister of Moffat'.[4] Indeed, Richardson's own sanity had been questioned, prompting him to sub-title a memoir 'with proofs … that the author is not mad'.[5] Such slurs often indicated lack of understanding, but Richardson's reviewer *did* understand. By slating his method of disseminating improving knowledge by addressing aristocrats, he was being held up as an anachronism. Nevertheless, why did fiorin generate such controversy? This question centres on the battle for public opinion by an eighteenth-century Irish improver trying to promote his ideas in the newly created United Kingdom.

The context of improvement

As Richardson did not confine his ambition to Ireland, it is necessary to set the wider context of improvement, rightly called the 'leitmotiv of Georgian Britain'. Improvement was a broad if amorphous topic. It comprehended material and moral progress as well as agriculture, facilitated utilitarian economic advance and allowed the socially ambitious to adopt the role of the improver.[6] The Scot Lord Kames's influential *Gentleman farmer* (1766) represented improvers as paternalistic patriots who harmonised private and public good.[7] Improvement operated through formal mechanisms of parliamentary grants and also voluntary associations like the Royal Society.[8] Kames presided over Edinburgh's Philosophical Society (1768–83), which he saw primarily as an agent of agricultural improvement, though it later became dominated by professionals and landed men with smaller holdings but wider interests.[9] Influential statesmen joined these bodies. There was governmental overlap in the Board of Agriculture and Internal Improvement, established by Pitt in 1793, which reflected Kames's ideas and was modelled on Edinburgh's Highland Society. With the Scot Sir John Sinclair as President and the famous English agriculturalist Arthur Young Secretary, this government-funded body aimed to improve agriculture nationally by encouraging landowners to promote scientific farming and 'excite a spirit of industry and experiment'. As its title suggests, the Board considered topics from new breeds to drainage.[10]

There was no unitary model of improvement. It varied with time and place in response to different influences and motivations. London was important as the seat of government and the site of major scientific institutions, but there was no simplistic rural–urban dichotomy and the fact that there was a strong sense of regional identity cautions against overstating the draw of the centre.[11] Broadly speaking, with agricultural improvement, England was seen as the model for elsewhere.

Scottish improvement was strongly influenced by an Enlightenment reverence for natural philosophy and aspirations for cultural knowledge as a civilising pursuit, as well as for reasons of economic utility.[12] There was improvement in the Scottish Lowlands during the eighteenth century, but Highland agriculture lagged behind. The Edinburgh Agricultural Society (established in 1790) addressed this and many enthusiastic improvers were Highland landowners, like Kames, who owned Blair Drummond in Perthshire, and Sir John Sinclair, whose Ulbster estate was in Caithness.[13] The Scottish universities which produced Adam Smith and David Hume were more intellectually dynamic than their English counterparts, and had greater economic accessibility for professional men.[14] This widening social range saw smaller landowners and professional men join the ranks of the improvers, like Richardson's acolyte, Patrick Miller of Dalswinton, who made his money in Edinburgh trade and banking. He built Dalswinton House in the fashionable neo-classical style, experimented with steam-powered craft, and planted trees for aesthetic effect.[15]

Improvement in eighteenth-century Ireland also drew on Enlightenment ideas and involved voluntary societies, parliament and individuals. 'Speaker' Foster spent heavily on improving his Collon estate in county Louth. He had connections to men like Sir Joseph Banks in England and key Scottish Enlightenment figures like George Dempster of Dunnichen and Sir John Sinclair, who were knowledgeable about the European *philosophes*. Foster was a guiding influence in the Dublin Society,[16] which began life in 1731 to improve agriculture and manufacturing. Its membership was elite and Protestant, like that of the Physico-Historical Society (1744), which gathered information on history, customs and natural history to provide information for improvers.[17] Scientific associationalism crossed national boundaries. The Dublin Society had links with the Highland Society and the Board of Agriculture over issues of mutual concern.[18]

Education, 'manners' and health also had associational manifestations. Dublin, like London, had a Society for the Reformation of Manners.[19] The Incorporated Society received a royal charter in 1734 and was given parliamentary money to establish schools to encourage Protestantism and English-type industriousness, while county infirmaries appeared after 1765.[20] Members of the Royal Irish Academy (established 1785) like Lord Charlemont and Richard Lovell Edgeworth were 'active patriots and economic improvers' and its *Transactions* reflected Enlightenment values, being dedicated to 'science, polite literature and antiquities'.[21] Local and county farming societies had been established since the mid-eighteenth century and had around 10,000 subscribers by 1802.[22] The Dublin Society reached out to provincial improvers, but with limited success, though some county surveys were published, modelled on Sinclair's Scottish initiative. Its other initiatives included a geological museum in 1792, Glasnevin Botanical Gardens in 1793–95 and professorships in botany, mineralogy, chemistry and veterinary science. In 1800 the Farming Society of Ireland assumed the Dublin Society's agricultural role, distributing premiums, promoting agricultural shows and encouraging local farming societies.[23]

Around a quarter of the Irish parliament's bills or heads of bills between 1692 and 1800 related to improvement, including canals, fisheries, collieries, turnpike trusts, bog drainage and the Linen Board (established 1711).[24] MPs were often members of the Dublin Society and Linen Board. Some 'patriot' MPs alleged jobbery in premium distribution, but others thought subsidies stimulated economic growth. 'Legislative independence' in 1782 saw increased encouragement of improvements, but often without adequate planning. Though provision was less certain from the 1760s, the last Irish parliament ensured that the Dublin Society received an annual provision of £10,000 for twenty years.[25]

If improvement was multifaceted, so were its underlying motivations. Some saw agricultural improvement as fashionable: the fifth Duke of Bedford's experiments at Woburn were the subject of emulation.[26] Others, like Lord Kames, were economic patriots: disinterested landowners whose improvements benefited themselves and the 'nation'. Many Scots, including Kames, were becoming less narrowly patriotic, affiliating themselves to 'northern Britain', an ideology of economic improvement orientated towards England.[27] Sir John Sinclair shunned chauvinistic Scottish patriotism, but was patriotic regarding Britain as a whole and in particular Caithness. Such patriotism was not nationalism, though national individuality persisted: the Highland Society maintained a Gaelic bard and the Welsh had a cultural revival. Richardson's supporter, Thomas Johnes of Ceredigion, combined his 'picturesque' estate's tourist potential with agricultural improvement.[28]

Motivations in Ireland included economic patriotism, religious proselytising and social advancement. Irish landowners once saw farming as artisanal. As early as the 1760s it was claimed that 'a man has a figure in his county in proportion to the improvements he makes' and by 1800 'the gentleman and the farmer [were] no longer incongruous terms'.[29] Irish improvers were prolific pamphleteers but, unlike Swift, most advocated political solutions to Ireland's problems within extant Anglo-Irish economic restrictions.[30] As in Britain, much literature was directed to parliament. Bishop Francis Hutchinson wrote pamphlets on banking, fisheries, poor relief, bog drainage and river navigation.[31] Many improvers were landed or benefited from agriculture through tithe, and they too experimented with new crops and farming techniques. Anglican clergymen were prominent and a unique Hibernian feature linked improvement with proselytising. Improvers were 'conquerors and colonisers' who saw indigenous Irish practices in agriculture, dress and diet as inferior to English or Scottish ways.[32] Bishop Hutchinson, a Dublin Society member, wanted to convert Catholics to encourage civility. Richardson's grandfather, the rector of Belturbet, sponsored printing of the New Testament in Irish. Foster's improvements in Louth were 'inextricably' linked to promoting Englishness and Protestantism.[33]

This landed improving ethos forms the background for Richardson's advocacy of fiorin grass and his battle for public opinion. As the aristocratic composition of metropolitan improving societies suggests, Richardson targeted elite opinion. The printed word was the principal means by which opinion was shaped. The pamphlet or printed personal memoir was a vehicle to reach this elite audience. However, the growth of a newspaper and periodical literature in the early nineteenth century represents a widening of opinion. Though it is difficult to be precise about periodical readership, it was considerably smaller than that for newspapers and would have been limited to the landed classes and some middle-class elements.[34] However, farmers of a lower social status were not interested in reading about scientific farming.

As late as the 1830s the 'Farmer's Series', published by Henry Brougham's Society for the Diffusion of Useful Knowledge, tried to inculcate new ideas by producing cheap and accessible publications. But this level of opinion remained convinced that they knew farming better than any savants and, as the Society's Secretary, Lord Althorp, put it in 1840, 'I am afraid our Farmers' Series is not very much taken by the farmers'.[35] Though this level of opinion was beyond the pale, the aristocratic element Richardson initially sought was diminishing as men from mercantile backgrounds became improvers. Though he never abandoned the pamphlet and memoir, he was forced to intervene in the periodical literature. The ebbs and flows of this struggle are worth following as it began immediately after the extinction of the Irish parliament. Richardson's modes of knowledge construction were shaped in pre-Union Ireland, but he operated in the new United Kingdom. Unlike earlier Irish improvers like Bishop Hutchinson, who sought to import English models, Richardson was convinced that Ireland could give the lead to Britain. But Richardson himself was mercurial. To imprison this will-o'-the-wisp within a static context might produce superficially plausible arguments, but ultimately lead into the mire of confusion. That he had only one topic of improvement is less historically important than the breadth of his ambitions. Recent historiography examines how scientific knowledge was constructed and disseminated, and has focused on the recipients as well as the creators of knowledge.[36] This approach informs the next section, which examines how Richardson constructed fiorin knowledge and tried to convince agricultural opinion.

Fiorin grass and aristocratic opinion in Ireland and Britain

Although he was not a member of the Dublin Society,[37] it was indirectly responsible for Richardson's first mention as an agriculturalist. This dates from July 1800, when the Society's surveyor of Tyrone was impressed by his experiments, which showed 'the spirit and talent for useful farming'.[38] Richardson thought so too and solicited the Earl of Meath's patronage, visiting his Kilruddery estate in 1806 and securing promises of cooperation in experimentation and a request for a personal memoir highlighting 'those [indigenous grasses] which promise to be the most useful'. Richardson played his cards carefully. While this memoir mentioned fiorin, its thrust was ideological: improving knowledge must be practical and not merely 'philosophical' or theoretical. Experimentation was not an end, but a means to a utilitarian goal. Parliamentary subsidies should encourage practical experiments but public money must not be squandered on botanical science. Farmers were baffled by the complexity of botanical classification, for example as was practised in Glasnevin Botanical Garden. Parliament's 'national liberality' had established the Garden under 'that most respectable body, the DUBLIN SOCIETY', but the money was wasted as the enterprise lacked 'the application of theory to practice'. Glasnevin's botanists listed many grasses but ignored their 'possible utility', giving the 'experimentalist' nothing to offer the 'practical farmer'. Proper patrons were not self-professed experts but patriotic noblemen like Lord Meath, who 'proved his attachment to Ireland, by his attention to her public business, her agriculture and her improvement'.[39]

Underlying this gentlemanly appeal were crucial differences about improvement's purpose and the relative merits of practicality or the botanists' more scientific and theoretical approach. This issue had long divided the Society between those, like Foster, who supported practicality, and others who were 'over-exercised with theory or pure science'. As a forum for elite improvers, the Society carried weight with wider opinion, hence Richardson's focus. Foster saw Glasnevin as an instrument of improvement by providing a training ground for the landed classes, which would help them become leaders in society.[40] Acting on his own initiative, he had ensured the appointment Walter Wade, a surgeon, as botany professor 'to encourage the study of botanical knowledge for the benefit of agriculture'. However, scientific botany seemed to be marginalising practicality. The Society was awash with cash, being funded by continuing parliamentary subsidies.[41] Substantial premiums and a gold medal were offered to whoever answered best on botany by public examination. Only those involved in 'practical' farming were supposed to be eligible, but the emphasis on 'botanic descriptions' disgusted Richardson, who saw this as public money wasted. As Lord Meath had been a member since 1801, Richardson's memoir was aimed at putting influential improving opinion back on the right ideological track, emphasising utility over theory.[42]

Richardson advocated fiorin as an improving agent, but its acceptance by agricultural opinion depended on wide recognition of its utilitarian properties for reclaiming wastes and producing winter hay. His confidence had been immeasurably boosted when Humphry Davy visited Portrush in July 1806. Davy found Richardson busy with an experimental fiorin plot, recognised its practical potential and asked him to write a memoir for the Board of Agriculture.[43] Davy was the Board's lecturer in agricultural chemistry and his influence ensured this memoir was well received. Richardson was awarded a medal and his memoir was printed in 1807. 'Calculated solely for an English audience', it claimed that fiorin would convert bog and heath into fertile pasture and meadow. 'Men of property, spirit and enterprise' were challenged 'to engage in the improvement of their estates on a gigantic scale'.[44] The Board's sway was extensive: the Dublin Society publicised its premiums, making it ideal for Richardson to access aristocratic opinion in Ireland and Britain.

The Board of Agriculture memoir omitted Irish information, but Richardson covered this in an 1807 memoir on 'flow bogs' to Ulster county grand juries. These bodies reflected elite landed society, so the same audience was targeted. This claimed that fiorin would improve the 'deep flow moss' peculiar to Ireland. The focus was on the 'Montiaghs', a 'dreary tract' of land around Lough Neagh, but the method was applicable elsewhere, especially the Bog of Allen. Practicality was stressed. Though philosophical naturalists speculated endlessly on how peat formed, the robust 'agriculturalist' seeking profit had 'utility as his sole object'. Simple experimentation contrasted with the narcissistic preoccupations of botanists fixated on discovering new varieties of grass but providing no useful knowledge. Unconsciously anticipating later arguments about natural selection, farmers should ignore useless books and 'consult nature herself', as plants were conditioned by environment. Fiorin flourished in Irish bogs, producing long 'strings' (stolons), which grew in all directions, forming a mat for cattle to graze upon. Under cultivation, the grass was not propagated from seed but by these strings, cut small and sprinkled with compost. A small plot would produce enough for inexpensive trials. 'Amateurs' were invited to inspect Richardson's experiments and they could provide sworn testimonies – an accepted means of knowledge construction for premium applicants.[45]

Richardson's geographical seclusion compromised his ability to convince elite British opinion. Clonfeacle's location discouraged physical inspection and the nature of experimentation meant that the knowledge soon dated. He published updates, including a memoir for the Belfast Literary Society adding another year's experimentation and making expanded claims. Fiorin flourished in rich or poor soil, grew on hills and valleys and occurred as far afield as India. Testimonies were included from landed and military men. Colonel Macan had commanded cavalry in India and believed that fiorin was like Indian 'Doob grass'. Fiorin could be harvested for hay in the winter and Davy's support meant the wild 'doctrine' that fiorin overturned the conventional wisdom of 'making hay while the sun shines' was legitimised as agricultural knowledge. Analysis suggested that fiorin hay was superior to its summer equivalent. Sensing he was regaining the initiative, Richardson attacked alternative knowledge constructs. Doubters called fiorin 'squitch' grass, a farmer's pest, but his experiments revealed the differences. New trials under irrigation, a 'modern improvement' involving diverting water along artificial channels to increase the land's productivity, promised further revelations.[46]

Claims about fiorin being a useless weed epitomised the opposition Richardson faced in Ireland and Britain. These serious challenges were grounded in utility, not botanical categorisation, and hence were harder to oppose. A prime objector was Arthur Young, an honorary member of the Dublin Society who advised on premiums.[47] His *Annals of agriculture* declared that livestock refused fiorin, which was actually the weed 'red robin'. Richardson countered by sending specimens which his own cows relished.[48] Young relayed these to Sir Joseph Banks, who deferred any 'opinion relative to the much controverted question of Fiorin Grass' without more evidence, but thought the cuttings were probably red robin, 'a most mischievous weed in corn'. Banks was keeper of Kew Gardens, London's equivalent to Glasnevin, and had a keen interest in botany.[49] He identified the grass under its Linnaean classification as *Agrostis stolonifera* but believed its stolon-producing property varied in different soils, leading botanists to think there were distinct species. Fiorin flourished in Ireland's 'damp western air' and in Devon, but only further experiments could establish its real utility.[50] Walter Wade agreed about divided opinion, saying fiorin was praised in Ulster but dismissed by farmers in England and Kilkenny as 'troublesome' couch grass or 'black squitch'. Nomenclature was also problematic. Attempting to reconcile colloquial English, Gaelic and Linnaean classifications, Glasnevin's under-gardener, John White, noted category confusion.[51] Thus, as Richardson's aim widened, he had to adjust his opinions and pragmatically tailor the means to the end.

Richardson's early publications reveal worries that his practical ideas might fall on stony ground. Many potential patrons were interested in scientific botany as well as agricultural improvement. So, despite his contempt for botanists, the nomenclature problem had to be addressed. To reach his goal of national improvement on a massive scale, he must first prove fiorin's improving credentials in Ireland, then demonstrate its ubiquity throughout the United Kingdom. Indeed, the entire project rested on establishing this beyond contradiction. He needed a champion with connections in both islands whose social status would elevate Richardson's protégé by generalising and validating the knowledge he was producing. His old political patron, the Marquis of Abercorn, was that man.

Abercorn was an enthusiastic improver who also had important Scottish and English estates, including Bentley Priory in Middlesex, near Harrow School, where he was a governor. He owned a London house at Grosvenor Square and in 1805 one of his daughters married the 4th Earl of Aberdeen, whom he helped into the Scottish peerage.[52] Richardson enlisted 'Magnifico' in the fiorin cause, sending a copy of the Belfast memoir and advising him to write to London for the out-of-print Board of Agriculture memoir, stressing that this indicated the Board's approval. Abercorn enthusiastically asked about fiorin cultivation. Richardson, acutely aware of the boost such patronage would give, directed his Dublin publisher to send Abercorn and his land steward, John Burgoyne, copies of the other memoirs. In June 1808 Richardson invited Burgoyne to view the Clonfeacle experiments, visited Abercorn's estate himself and, though almost seventy, led a party up Bessy Bell mountain, to declare triumphantly fiorin culture to be feasible.[53] Abercorn's doubts over expense were removed as Richardson stressed the economic and patriotic benefits. Ulster was like the classical 'officina gentium' (the workshop of the world) and its population was rising rapidly. Fiorin could help feed them by converting useless land into meadow, allowing cattle to move from lands which could be turned to tillage. Burgoyne suggested a prize for the best crop.[54] Abercorn seemed willing and Richardson, concerned that the Board of Agriculture memoir had dated, published revised material with additional testimonies.[55]

Abercorn was valuable in the nomenclature and ubiquity struggle, and received specimens to enable his English 'botanical friends' to ascertain whether fiorin was identical to 'Orchestron grass'. Having solicited landowners' heroic efforts, Richardson now used an imperial metaphor, equating reclamation with conquest and quoting Lucan's epic poem about Caesar, *nil actum credens cum si quid superset agendum* (think nothing done if anything remains to be done). Fiorin promised Abercorn the 'annual and rapid improvement of his country' and inexpensive political and cultural kudos. Abercorn's example would encourage lesser landowners, who could improve without increasing rents yet also benefit 'the nation' by making huge wastes profitable.[56]

Concerned about fiorin's ubiquity, Richardson also exchanged specimens with Lords Redesdale and Rous and sent notes to England's 'great patron of agriculture', the Duke of Bedford.[57] Banks received fiorin from Richardson, urged its cultivation in England and forwarded it to Sir James Smith for experimental planting.[58] Best of all, despite Arthur Young's objections, matters looked promising with the Board of Agriculture, a key influence on aristocratic opinion. Richardson's medal-winning memoir followed Sir John Sinclair's re-election as President of the Board in 1806 and he recommended fiorin experiments in his 1808 presidential address.[59] Therefore, two years after Davy's visit, the campaign looked most promising. Fiorin seemed to be ubiquitous in Britain and Ireland. Abercorn, Bedford and Banks were all experimenting. Sinclair found fiorin growing naturally in Scotland and was 'laying down an acre close to the town [Edinburgh], to teach his countrymen'.[60] In October 1809 he wrote to Richardson 'in the name of several members of the Board of Agriculture', praising his patriotic service to the country and promising, 'if the plan succeeds', a parliamentary motion for a grant of £1,000.[61]

This scenario would have ideally fitted Richardson's notion of how improvement should be organised. Government would facilitate and encourage while landowners would determine what improvements were suitable for their estates and their implementation. This presupposed strategic governmental intervention but, in 1809, it seemed public money would be thrown at the problem of unproductive wastelands through the Irish Bogs Commission, an idea of John Foster's. Richardson's complex ideological reaction will be discussed in the concluding chapter but, by funding surveys conducted by experts susceptible to the influence of jobbing landowners, the Commission threatened to corrupt agricultural opinion and hence greatly increased Richardson's urgency to convince aristocratic landowners.

The personal memoir was an accepted way of disseminating knowledge among the elite and Richardson addressed one to the Earl of Rosse in April 1809. As Viscount Oxmanstown (his designation before assuming the Rosse title in 1807), he had supported Foster, but their ways had parted in 1806 as Oxmanstown retained his Treasury office under the Talents government when Foster had lost his position.[62] Richardson's memoir to Rosse outlined plans for reclaiming the Bog of Allen with fiorin. It included testimonies from 'the first noblemen and gentlemen of the country' and an epitome of fiorin's 'habits, properties and uses' for reclaiming different types of bog, reprinted from the *Gentleman's Magazine*.[63] Richardson boasted that Rosse was 'in league with him' against 'the old brown gentleman' (the Bog of Allen) and noted Foster's 'tirade against the same bog' in the English papers. Rosse was the first nobleman to undertake the challenge to reclaim unproductive land on a 'gigantic scale'.[64] Richardson followed his written words with a personal visit and explored the Bog of Allen with Rosse in June 1809.[65]

The battle for Scottish opinion also continued apace. Richardson courted Abercorn's son-in-law and Highland Society member, the Earl of Aberdeen, who had found fiorin on his estate and sent his agriculturist to Clonfeacle to learn about its cultivation.[66] Aberdeen's influence made his support crucial and urgency informs Richardson's correspondence with him. On receiving a letter as he left Portrush, he did not wait to reach Clonfeacle but penned a hurried response at an inn, boasting that fiorin had 'risen infinitely in esteem'. He mobilised Irish knowledge for a Scottish context, directing Aberdeen to his Dublin publisher for the Bog of Allen memoir to show how fiorin worked on both wet bogs and dry heathy moorland. Opportunities beckoned to influence other Scottish aristocrats. The Highland Society's *Transactions* reported only moderate productivity from the Duke of Buccleuch's water meadows, but Richardson claimed that produce would treble if fiorin replaced 'paltry grasses'. Desperate to make inroads in Scotland, Richardson sent Aberdeen instructions about fiorin cultivation, as 'a failure … would be serious and we must take care to prevent it'.[67]

Richardson's bid for Scottish opinion also involved extensive interchange of fiorin specimens, which were posted from as far afield as the Shetlands. Scottish visitors also came to Ulster, including the Earl of Selkirk, who visited Richardson in 1809 and took a sample of fiorin back to Kirkudbright and was thus able to show that the same grass was abundant locally. Another visitor, the Edinburgh banker Thomas Allen, confirmed that 'fiorin experiments are in many hands and the grass itself everywhere spontaneous in the vicinity of that city'.[68] The Edinburgh lawyer and upwardly mobile landowner Alexander Young of Harborne got samples of winter-mown fiorin hay.[69] Patrick Miller began planting fiorin in April 1809, aiming to increase his rental by £3,000–£4,000.[70] Miller converted thirty acres of waste to valuable land which could be let at £20 per acre, prompting Richardson's comment that he was 'the most zealous and most extensive cultivator of fiorin that has yet appeared'.[71]

This engagement with Scots like Young and Miller who had moved from mercantile or professional backgrounds into landed society had profound consequences for Richardson's campaign for public opinion. He needed to address a wider social range of opinion; but this had to be pointed out. As support grew, suggestions were made that he must extend his publishing to increase publicity. Eager to establish a foothold in Scotland, Richardson's instinct was to increase pugnacity, which proved dangerous. Richardson bragged that Aberdeen could 'mow a crop of hay superior in quality and double the quantity of any in England' and he vowed to 'have a brush' with any contributor to the Highland Society's *Transactions* who dared 'abuse my protégé fiorin'.[72] More level heads like Thomas Allen persuaded Richardson to drop his belligerence and court opinion by contributing to the Edinburgh-printed *Farmer's Magazine*. Though not having a mass circulation, such periodicals reached an audience which included landed and professional men.[73] Intervening in periodicals allowed Richardson to field enquiries and to position his knowledge to differentiate between Scotland and England.[74] Solid Scottish support could help fiorin cross the border into England. Though periodicals expanded the audience, increased publicity came at a price.

Richardson also began publishing in the London-printed *Agricultural Magazine* and here he began to encounter opposition. A Surrey farmer had 'magnificent accounts' of fiorin from a member of an Irish agricultural society, but was sceptical as his father had lost a fortune trying Jamaican 'Guinea Grass'. Conscious of the pun, he accused naive persons of promoting 'foreign' grasses, assuming they would flourish everywhere.[75] Even in Scotland objections were raised. Selkirk was 'pronounced mad' for ordering land to be 'laid down with Squitch Grass, for so they denominate my fiorin', instead of wheat.[76] Another Caledonian sceptic demanded the Linnaean term for fiorin, claiming that if it was *Agrostis stolonifera*, then it was a 'troublesome weed … called black squitch', useless on arable ground.[77]

This uneven reception was unsettling for Richardson, who had previously relied upon normal eighteenth-century modes like private publication and gratis distribution to individuals.[78] He usually published 'at his own suite' or following solicitation by a landed proprietor for a memoir which the gentleman printed at his own expense and distributed to friends. These traditional methods had the benefit of keeping control of the knowledge firmly in the hands of its originator. This was not so with commercial publishing. His own family history seemed to prove its incompatibility with landed interests. His grandfather, John Richardson, seeking to print Bibles in Irish, had fallen into the clutches of mercenary

London booksellers, who forced the sale of an estate to pay the debt. Yet, on the other hand, for someone with national ambitions, private publication also had weaknesses. It relied on personal contact and tended to repeat the same information. Richardson reluctantly realised that still more was required to convince wider opinion. In November 1809 he had told Greenough that he 'considered my best policy under Sir John Sinclair's kind intentions, was to put forward … whatever important *untouched* matter remained with me on the fiorin subject'. Marooned in Ulster, Richardson relied on Greenough for information about 'Sinclair and the fiorin question' and asked him to forward immediately 'anything hostile' in print on the subject. The new subject of irrigation would generate publicity. His 'name may be obscure now' but 'the incalculable importance of my discovery' would change that. Speed of response was crucial, but Irish printing had 'no vigour, even in Dublin' and was too slow 'for the present emergency', so his irrigation pamphlet went to a London bookseller.[79]

Forsaking the landed gentleman's experimental meadow for the commercial marketplace was painful. The logic of his ambitions meant that Richardson's agricultural knowledge had to become a commodity, not a gift. It was no longer an unmediated transaction between like-minded men which flowed as easily as port after dinner. Now, in the cold light of day, knowledge had to be purchased and be deemed worth buying. It was a transition Richardson never fully made. Reluctantly accepting market demands, he remained uncomfortable with commercial publishing, wanting its benefits yet hankering after the old ways of personal contact with the elite.

In response to Greenough's well intentioned suggestion about circulating his ideas to Woburn, he complained about losing personal contact. Woburn was not 'accessible to me', while some 'amateur' conveying the essays to Bedford was just not the same.[80] Resort to the market risked alienating elite patronage and Richardson tried to reconcile respectable and commercial publicity. He let Aberdeen print 'an elaborate letter on the improvement of the Highlands with some wandering into England', and felt 'highly honoured by being permitted to go into the World in such company'. Yet, as his overworked amanuensis struggled to keep up, corners were cut. Though Richardson still disavowed 'booksellers and their claims' the 'letter' was a pro forma: 'instead of making queries, suppose your lordship states queries *to me*; leaving in your letter blanks for my answers', thus saving time 'in my present pressure'.[81]

Increased publicity had implications for control over knowledge, as Richardson's carefully cultivated personal correspondence network was overwhelmed. Inundated by 'inferior applicants', he sought a socially constructed filter in order that he might maintain personal connection with his most important contacts. Priority was given to 'old correspondents' like Aberdeen and correspondence continued with elite improvers like the Somerset MP William Dickinson. Richardson also tried old methods for new media, getting Greenough to circulate offprints from *Farmer's Magazine* to fellow MPs.[82] He intended to influence legislators in a way familiar to eighteenth-century writers of improving literature.[83] But public exposure and open criticism increased in ratio.

In Ireland, despite his criticisms of the Dublin Society, Richardson sought one of its premiums. He submitted samples of fiorin hay with testimonies from his steward, a land surveyor and his old Trinity pupil and Dublin Society member Isaac Corry, who had overseen large government grants to the Society during his period as Irish Chancellor (1799–1804). Although Richardson's attempt got no further than a conventionally polite notice in the minutes, his efforts for publicity earned public scorn in another medium.[84] A Dublin-printed caricature entitled 'Rich-en-son grass, or making hay when the sun *don't* shine' depicted him as manically reaping in pelting rain beside a pile of publications. This print, addressed to 'all cultivators of Fiorin Grass in the United Kingdom', aimed to stymie attempts to promote the grass outside Ireland.[85] It is not clear who the instigators were, but the implications were that Richardson was mad. His immediate response was 'a small squib' addressed to Corry intended 'to stop the mouths of my London friends who are getting very importunate for the defence of my sanity'. The critical stage of his battle for public opinion made Richardson opt for a Belfast publisher, though he complained about delays. Aimed at an English audience, when this piece eventually emerged, it outlined fiorin's properties to prove 'that the author is not mad'.[86] Richardson resolutely defended his counter-intuitive approach: 'Christmas hay-making is so contrary to experience, so revolting … that the improbability of success is … an excuse for declining … experiment'. At Trinity he had taught Corry the principles of demonstration: *a priori*, showing what ought to be found by the nature of the subject, and *a posteriori*, proving empirically that the facts corresponded. Respectable witnesses confirmed six tons of hay per English acre (rather less than his Dublin Society bid, which claimed eight tons). The argument was a mixture of patriotic rhetoric and economic logic. Summer haymaking's advocates enjoyed an 'unreasonable monopoly' until confronted by empiricism's 'logical weapons'. Would his critics, the 'academic gentry', prefer lands to stay waste? 'The state itself', facing population growth, should promote fiorin and support landowners to create additional pastoral land and free existing grasslands for tillage.[87]

Any publicity was good publicity. As the controversy grew, Richardson's friends urged more exposure. Greenough suggested that a 'collection' of essays 'ought to bring money from a bookseller'. Richardson, tempted, asked his friend to use his influence with Phillips, the Board of Agriculture's London printer.[88] Yet he still perversely persisted with traditional modes. County Down's largest landowner, the Marquis of Downshire, got an unsolicited letter ostentatiously mentioning notable fiorin practitioners like Lord O'Neill, Corry and Banks. Downshire, like Abercorn, heard how 'cut-out' bogs could be reclaimed to the benefit of landlord and tenant. If major proprietors gave the lead, the inhabitants would undertake this improvement by planting fiorin on bog-land. Downshire's interest allowed Richardson to walk the well worn path to the aristocrat's mansion and follow his initial letter with one mentioning irrigation and enclosing his 'letter' to Corry. However, other aristocrats recognised Richardson's need to break free from gentlemanly publishing.[89]

Lord Hertford was 'a great Amateur of Fiorin' with estates in Ireland and England, where he found it resisted severe drought. Like Greenough, Hertford saw the need to reach a wider public by publishing all privately printed memoirs in an agricultural journal. He even offered to print Richardson's work at his own expense and distribute it to MPs. Concerned for his social reputation, Richardson ignored this and sent Hertford a personal memoir arguing 'that all wastes, Fens, and heaths in England may be clothed with this grass'. Stung by criticisms about winter haymaking, he asked 'would his lordship also be considered mad?' Surely Hertford's experiments, corroborated by testimony from Banks and several Irish peers, meant that his respectability would counteract the ridicule.[90]

Publicity also exercised a Scottish visitor to Clonfeacle. Patrick Miller of Dalswinton sent his land steward, John Farish, to learn fiorin cultivation. Farish was highly impressed and sent the *Belfast News-Letter* 'a very smart puff', explaining how Miller, well known in Ireland, had been induced to try fiorin. Farish urged 'universal publicity' as 'many of our nobility and gentry in the United Kingdom are now giving a laudable attention to national improvements' like getting hay from the 'unprofitable muirs [*sic*] and disgusting morasses of Scotland and Ireland'.[91] Farish recognised Richardson's limitations and, to help publicity, produced his own fiorin 'treatise', printed by Constable, publisher of *Farmer's Magazine*, and sold in Dumfries, London and Edinburgh. This compared Richardson's 'public spirit' to Miller's patriotism for promoting public as well as private improvements. Fiorin grew best on 'low damp soil' but worked elsewhere, promising 'immense individual, even national advantages'. Only Richardson had found the grass to succeed under irrigation and Farish recommended the irrigation pamphlet, 'if Dr Richardson allows it to be publicly sold'. Farish's appendix rubbished accusations that fiorin grew only in Ireland and was 'vile trash', unpalatable for animals.[92]

Richardson knew he needed publicity in the struggle for opinion, but was hampered by distaste for publishers' advertisements, hoping it would 'not look as if I had any concern in the profits'. He also asked Phillips to distribute complimentary copies to noble correspondents 'at the request of the author'. This was ignored and Richardson complained that 'I hate booksellers' and felt 'entitled to more [complimentary copies] for my friends in Ireland, otherwise it will never reach this country'. He also baulked at a compendium. Gentlemanly publishing had the benefit of serialisation, as each essay was distinct, but a collection would repeat things already in print. Richardson naively hoped that Phillips would deal with the non-aristocratic correspondents, but he refused and eventually ended contact with his demanding author.[93]

Sir John Sinclair's backing would have put fiorin back on the aristocratic trajectory Richardson had originally envisaged for it.[94] But by mid-1810 he complained that 'I know not what to make of Sir John Sinclair', and was 'nettled' at his 'contemptuous mention of fiorin in his farewell speech' to the Board of Agriculture. Richardson thought ubiquity was the issue and vowed to 'immediately *sustain* … that there is not an acre in England which may not be made productive … by the aid of Fiorin Grass'.[95] In 1811 Richardson was hurt by Sinclair's 'cold incidental mention of Fiorin' when the subject was broached in London. Worse would follow. When Sinclair gave fiorin to Mr Salisbury of the Highland Society, he exploded, 'Damn it altogether, it is nothing but squitch!'[96]

Despite these setbacks, the battle for public opinion was not yet lost. *Farmer's Magazine* published a letter from Richardson's Edinburgh correspondent, the advocate John Sheriff, whose experiments seemed to disprove that fiorin was a weed. He promised colour plates for easy identification and dismissed sneering anti-Irish references in English publications like the *British Critic*, which asserted that fiorin-inflated land values would replenish absentees' wine cellars depleted by Union. His fellow advocate, Alexander Young, reckoned fiorin was excellent for feeding to ewes and General Dirom of Dumfriesshire found it good fodder for horses. Publishing coloured plates was exceptional, but the high interest in fiorin justified it. The August edition of *Farmer's Magazine* was graced by two beautifully coloured fold-out plates, accompanied by Sheriff's debunking of claims published by the Highland Society that livestock refused fiorin. Richardson tried to exploit tensions in that Society, claiming 'Gentlemen agriculturalists', like Sir James Stewart of Coltness, were naturally enthusiastic, without any premiums. Fiorin's 'real enemies' were seed merchants, jealous as fiorin did not grow from seed, and land stewards, who misadvised nervous landed proprietors afraid of anything new.[97] Plenty of respectable opinion could be marshalled against Scottish setbacks, but England was the real problem. Richardson knew this and utilised the provincial press. Fiorin's economic advantages could be seen in their full light at Chester hay market. The *Lancaster Gazette* printed a letter by 'Publicus' which echoed Richardson's view that innovations usually met with contempt from the ignorant, but 'enlightened' men like J. C. Curwen MP, who was initially sceptical, now recognised fiorin's 'vast importance'. Further letters endorsed this view and referred to the regiment of respectable testifiers Richardson mobilised in his publications. The *Gazette* even printed a coat-trailing letter by Richardson to the *Chester Chronicle*. 'John Bull I know is very wise', he wrote provocatively, 'but when he sees the produce of his rich meadows undersold [at Chester market] from Welsh mountains, Irish bogs and Galloway moors … we shall have a fair contest – the Aborigines against the Saxons'.[98]

Richardson's private correspondence reveals similar attempts to embarrass landed men into compliance by claiming elite support and disavowing the periodical literature he himself had to embrace. Brigadier-Major Kater of Ipswich heard that generals, attorneys, parsons and 'head clerks in the great London offices with country boxes' were all 'running wild for fiorin', but 'the agriculturalists are hostile, the higher the rank the more inveterate'. Having discovered a valuable grass which grew unnoticed before his opponents' eyes, Richardson gallantly refused to fight the ignorant, 'except when I know them to be mere hirelings of agricultural magazines and newspapers'. 'Never fear', he bragged, 'Fiorin will make its way, for when we have the *Church*, the *Army*, and the *Law* with us, who cares for the farmers, and still less for the vain and dogged land stewards'.[99]

Though he used the press himself, Richardson still worried that newspaper debates stimulated landowners' instinctive conservatism and it would be better to address them directly. Regional agricultural societies were an option. They were supported by the elite but also included working farmers and businessmen. The Middleward of Lanarkshire Farmers' Club was patronised by a Highland Society member, the Duke of Hamilton. Its founders included one marquis, one earl, three lords, one baron, one colonel and six esquires, and Richardson's supporters, Stewart of Coltness and John Baird of Shotts Iron Works in Lanarkshire.[100] The Kirkcudbrightshire Agricultural Society, which Richardson visited, was also highly respectable, meeting in Castle Douglas and having the Marquis of Queensbury and Sir Alexander Gordon as members.[101] This social mix in agricultural societies meant that they included wider agricultural opinion but the old elite still played a leading role. Richardson saw them as important enough to justify a visit to Scotland and Cumberland.

Agricultural societies could also help knowledge ramify more widely. His Edinburgh supporter, Robert Ainslie, lectured on fiorin to Kirkcudbrightshire Agricultural Society, which published the lecture in April 1811. Richardson's printed knowledge was thus transformed into oral information and then back again into print in another form. These Societies were also embedded in social and professional networks, and Ainslie knew of fiorin from his business partner, Stewart of Coltness. Societies could help to legitimise agricultural knowledge. Ainslie noted that though fiorin innovations risked prejudice, its widespread cultivation could be justified on Malthusian and theological grounds. Because of smallpox inoculation, population growth was inevitable and would outrun the food supply, but Providence had a remedy in nature which it was man's task to discover.[102]

Workington Agricultural Society was another associational stalwart, considered the 'most respectable' of its kind in England. Richardson was invited to a dinner attended by 700 people at which the President, J. C. Curwen, publicly stated his conversion to fiorin as a result of Richardson's utilitarian arguments.[103] Curwen's address focused on lowland Scotland, where domestic industry and population growth necessitated good yields from small acreages, and he mentioned the support of Dr Walker, Professor of Natural History at Edinburgh. Richardson toured Cumberland with his 'kind and intelligent friend' Curwen, an agricultural improver as well as a politician.[104] This Society had an economic hinterland 'across the firth'. It reprinted Patrick Miller's letter to the *Dumfries Journal* which noted his erection of 'Clonfeacle Tower' ('Miller's Worm' to his tenants) with an inscription praising Richardson as the patriot 'who was the first to teach others the culture and uses of this invaluable gift of a beneficent PROVIDENCE'.[105] It offered a premium to encourage fiorin's adoption in England. So important was this associational support that Richardson risked the Primate's wrath for neglecting his parish responsibilities by visiting Workington and Wigtown Agricultural Societies. The risk was worth taking. Richardson heard Curwen 'recant' his fiorin scepticism and received a presentation cup and a request to contribute to the Workington Society's anniversary proceedings.[106] As well as facilitating publication, societies could also validate experimentation. Clonfeacle was successfully inspected by the committees of two Scottish and one Welsh society.[107] *Farmer's Magazine* reported the findings of the Kirkcudbrightshire's inspection and reprinted Ainslie's *Memoir*, with a postscript justifying its inclusion because of 'the importance of the subject'.[108]

For a while this strategy of harnessing the social networks and disseminating prowess of agricultural associationalism seemed to be working, even in England. The Earl of Hardwicke was President of Cambridgeshire Agricultural Society, which offered a fiorin premium.[109] Support like this fed back into the pages of periodicals. For every critic in the *Agricultural Magazine*, like 'Cultivator Middlesexiensis', who denied fiorin's ubiquity, there was a balancing positive comment. A review considered Farish's 'very sensible' fiorin pamphlet as redolent of Miller, an 'eminent improver of that enlightened part of the kingdom'. Another contributor even demanded that Arthur Young reconsider his denigration of fiorin. The editor, Vaughan Griffiths, allowed Richardson latitude as he sought 'neither fame nor profit' and had been 'scandalously attacked' by those who should know better, including a Linnaean Society member who persisted in calling fiorin 'quitch'.[110]

Periodicals functioned as an early nineteenth-century variant of the classic Enlightenment notion of the republic of letters, by reprinting material from other journals. The *Weekly Entertainer* carried a lively letter from *Farmer's Magazine* in which Richardson had asked why, when fiorin had succeeded in Ireland and Scotland, the English could not 'conquer old prejudices'. In the United States, Enos Bronson's *Selected Reviews of Literature* included fiorin in excerpts from foreign magazines on subjects ranging from ballooning to Walter Scott's poetry. Published in Philadelphia and sold in New York, Boston, Richmond, Augusta, Savannah and Lexington, Bronson's periodical noted that fiorin had 'excited so much interest' in Britain and Ireland, it merited the inclusion of a line drawing.[111]

Richardson also tried to mobilise older contacts. Wartime restrictions notwithstanding, he solicited support beyond the British Isles. Ex-President of the United States Thomas Jefferson asked a Belfast contact for specimens and a copy of Richardson's Belfast Literary Society memoir, believing fiorin to be of 'inestimable' value on American 'galled lands'.[112] The Copenhagen merchant Frederick de Conynck also asked for fiorin information. Despite communication problems, Richardson sent stolons via Liverpool and Harwich, with copies of memoirs, and cleverly had his covering letter printed in the *Belfast News-Letter* on the pretext that it might miscarry due to the war.[113] Richardson also drafted a letter to his old geologist friend Professor Pictet, recommending fiorin for French mountains and bogs, noting that 'though we are engaged in war, the sciences are at peace'.[114] Thus Richardson campaigned very widely to convince agricultural opinion and, without parliamentary or Board of Agriculture support, used every publicity mechanism in his power.

Fiorin grass and agricultural societies

Yet Richardson's difficulties mounted as he oscillated between courting aristocratic opinion and a more socially mixed audience. Neither approach was ultimately successful, but each provoked revealing debate. In 1811 he had had success with the Farming Society of Ireland. Spawned from the Dublin Society, it was the Board of Agriculture's Irish equivalent, and its committee members could attend the British body. John Foster linked the two organisations, being President of the Farming Society and a Vice President of the Board of Agriculture.[115] Richardson disliked Foster personally, but he was not deterred from seeking Farming Society support, possibly because this organisation was still, until chartered in 1815, a voluntary body funded by subscription. It offered a gold medal for the conductor of 'an experiment which shall be the most satisfactory in establishing its [fiorin's] utility and merit'. Richardson saw a way to counteract English criticisms and addressed a memoir to the Society, which was published in London priced at 2*s* 6*d*.[116] He still focused on convincing the government of the need to support fiorin, describing how its hay survived snow and bitter winds to produce good feed. Diplomatic for once, he speculated that the prejudice he encountered might only be semantic. People associated the term 'hay' with summertime; but, though 'stolones' was an accurate description of fiorin hay, farmers would not understand it. Testimonies of success were laid on thick and fast. Enquiries came from the West Indies, Portugal, Philadelphia, Georgia and Denmark. Richardson's and Alexander Young's experience showed that sheep and horses preferred it, fiorin-fed cows gave better milk and the Wicklow mountains promised to supply Dublin with hay. Patrick Miller's and Robert Ainslie's testimonies corroborated fiorin's ability to thrive while submerged in water. 'Inveterate prejudice' that fiorin was a 'vile weed' had no national origins: English and Scottish agricultural writers were equally guilty of contempt before investigation. The Bath and West of England Agricultural Society's offer of a premium for fiorin trials on Dartmoor was held up against English naysayers. Scotland's agricultural writers were distinguished from its 'intelligent inhabitants', who would not reject anything without trial. Fiorin faced less opposition in Ireland, where the 'native Irish' knew of it despite being ignorant of cultivation. An inspection of Clonfeacle would establish fiorin's 'merit and utility'. Farming Society members were 'too respectable to have their testimony doubted' and 'the world will acquiesce in your decision'.[117]

Richardson won the gold medal and in 1812 tried another memoir to the Farming Society, using its sponsorship of a report on Wicklow by the Reverend Thomas Radcliffe as a pretext. Wicklow was a sheep-rearing region and, as the Napoleonic war stopped wool supplies from Spain, this was nationally important. New breeds were vital but the harsh mountain environment made feeding difficult and Radcliffe had recommended fiorin. Richardson latched on to this, citing his experiments to prove that it grew on peaty mountains. The quality of its hay had been witnessed by an Irish MP, a general and the Kirkcudbrightshire Agricultural Society. But, with an Irish audience now in mind, he was again drawn into national rivalries, asking why fiorin should improve Cumberland, Cardiganshire and Galloway but not Wicklow. English writers were 'vain' about irrigation and 'boast much of the improvement their country has received from it', but forgot about the properties of vegetables grown under irrigated land. Of three grasses which grew in these conditions, only fiorin was useful. Plantations were also a feature of improvement, and one which Foster specialised in at Collon.[118] Fiorin cultivation did not preclude planting. The 'celebrated' Irish oak was 'fast disappearing' but could be re-established in plantations where fiorin would also thrive, giving thirty years of good meadow as the timber matured. 'Wise *English* instructors' were wrong, as fiorin, unlike 'squitch grass', would not inhibit the growth of young trees.[119]

The difference in emphasis between the two Farming Society memoirs suggests that the first was unsuccessful in convincing English opinion, while the second's more nationalistic tone reflects Richardson falling back on Ireland. But even here problems arose, particularly in the south. Munster's Cork Farming Society had begun under the Dublin Society's aegis in 1789. Since 1810 it had broadened its social base, merging with the Cork Institution and spawning baronial associations. It published the *Munster Farmer's Magazine*, which had a circulation of 1,200 and was taken by landed improvers to circulate among working farmers.[120] It should have suited Richardson; for example, it had reprinted Patrick Miller's letter and Ainslie's Kirkcudbright memoir.[121] However, the fact that such periodicals were forums for debate was dangerous.

The Cork Society's links to working farmers made the publication of any critical observations especially damaging. Lord Shannon's land agent, Horatio Townsend, tackled Richardson in the *Munster Farmer's Magazine*. Townsend had much practical experience and was praised for his Dublin Society-commissioned agricultural survey of Cork by, among others, the influential English agriculturalist Edward Wakefield.[122] Townsend damned fiorin with very faint praise, saying it might do some good in autumn and winter, but that spring and summer crops of rye and clover were much better. Richardson rushed to fiorin's defence, writing that agents and agriculturalists exerted undue influence on their landed employers, who would have been amenable if approached directly. But Townsend was formidable. He noted perceptively how Richardson's zeal made him overrate fiorin's benefits but ignore its deficiencies. Floundering in this war of facts, Richardson retaliated by personalising the issue, claiming that Townsend was a sycophant who mouthed the couch grass charge. But Townsend attacked the 'the reverend and respectable panegyrist of fiorin' on utilitarian grounds, denying fiorin's value as fodder. Eventually the editor closed the debate, leaving Townsend the high ground.[123]

In England Richardson concentrated on locations where environmental conditions were like Scotland and Ireland. Dartmoor had potential as a large area capable of improvement. It was part of the royal estate and the Prince of Wales had offered a premium. The royal imprimatur might carry more weight than all his pamphlets, memoirs and periodical contributions, and Richardson hoped that trying fiorin on Dartmoor, along with the Somerset MP William Dickinson's offer of experiments in his West Indian plantations, might finally sway opinion, particularly Sir John Sinclair's. Eyeing the Dartmoor premium, Richardson prepared a memoir for the Bath and West of England Agricultural Society and considered other areas, like the Cambridgeshire fens, which, he joked, not entirely appropriately, would surrender to his 'vegetable Bonaparte'.[124] However, the Bath Society's response typifies Richardson's difficulties in converting Irish knowledge into English practice. His memoir was intended for inclusion in the Society's *Annals* but, after considerable hesitation, was printed separately. The publication committee knew that the high interest in fiorin precluded delay; yet the topic was so controversial that they withheld official blessing, noting how Richardson's zeal provoked 'deep-rooted prejudices'. The Society offered a premium but its conditional terms epitomised Richardson's English problems. The prize was offered to whoever could establish fiorin's identity, characteristics, produce and uses and ascertain how these compared with other grasses in 'southern Britain'. Richardson knew a poisoned chalice when he saw it, and insisted that any English farmers trying for the premium should get their fiorin directly from him. He arrogantly contrasted ignorant agricultural writers with poets who studied nature, not names. Sir Walter Scott's couplet 'the knot grass fetters here the hand / that once could burst an iron band' showed he knew fiorin's properties (Scott actually called fiorin a humbug on visiting Richardson in 1814). Worse followed as Richardson claimed that the bard of the celebrated Welsh chieftain Owen Glendower had praised the grass, meaning that if Scots, Welsh and Irish knew of it, why not the English?[125] Owen Pughe, compiler of a Welsh language dictionary, had told Richardson of the philological affinity between the Irish Gaelic 'fiorin' and the Welsh *gwyran*, meaning 'ever-living grass', which Glendower's bard praised.[126] But fourteenth-century Wales was not nineteenth-century Bath and the reaction can be imagined.

Richardson was an honorary and corresponding member of the Bath and West of England Agricultural Society and his memoir appeared with members' queries about fiorin's productivity. He addressed a letter to the Vice President, Thomas Graeme, demanding to know whether any member had actually seen or tried the grass and accused Graeme of elevating theory over practice by denying that fiorin occurred on dry heath soil.[127] Further acrimonious correspondence with another member, Thomas Tanner, made its way into the *Agricultural Magazine*. Tanner had denied fiorin's ubiquity and questioned claims about its hay. Richardson replied that Tanner had confused fiorin stolons with the culmi of other grasses, and blamed writers like Davies, the 'statistical oracle of Wiltshire', who thought fiorin was couch grass. But Richardson went too far in trying to blacken Tanner's character, claiming that his focus on generative culmi recalled Linnaeus's system, whereas he, Richardson, had 'no desire to draw back the curtain to peep into the mysteries of nature'. Botanical voyeurism aside, if Tanner doubted fiorin's productivity he should try to convince Lord Hertford. But this unsavoury episode epitomises Richardson's loss of control over fiorin knowledge, glaringly exposed when Tanner closed the discussion.[128]

Richardson also courted controversy in Scotland. *Farmer's Magazine* carried letters on irrigation and the productivity of peat soil aimed at contributors to the Highland Society's *Transactions*, who, influenced by Dr Rennie, believed peat was sterile. Richardson's Irish experiments had disproved this, and he threatened to return to Galloway to demonstrate how peat bogs could be improved. Cultural affinity between Scotland and Ireland could be exploited when the opportunity arose. 'Atlanticus' extolled fiorin's value in the Hebrides. Though instinctively sceptical about new grasses, he was convinced by Richardson's 'respectable character' and his privileging of hard facts over airy speculations in a 'plain, unaffected, and confident manner'. He also claimed that Irish fiorin was called *Olach* in Hebridean Gaelic and an old Scottish proverb noted how horses delighted in it. A further letter from Kirkcudbright Agricultural Society reiterated positive impressions of fiorin's productivity for hay compared to a similar Scottish meadow, and animals' readiness to eat it. Richardson responded with a letter of thanks showing his standing with the Galloway gentry who received him in their houses and transported him in their carriages.[129]

The regional nature of agricultural societies allowed Richardson to show how fiorin could benefit entrepreneurs in improving contexts other than reclamation. John Baird of Shotts Iron Works near Glasgow was a hard-headed businessman, sceptical until he met 'the father of the fiorin' and heard his 'sound arguments' on utility. Rye grass would not grow in his cold district and hay had to be bought in, but fiorin flourished and Baird's horses and cattle liked its hay. Richardson instructed Baird's workers how to lay down fiorin using coal ash from their houses as fertiliser.[130] Similar pragmatic diversification is evident in Richardson's private correspondence. He assured Alexander Young that fiorin's advance meant the lucrative Edinburgh hay market beckoned.[131] But straws of disapproval also blew in the Scottish wind. Appearance in periodicals was no longer automatic. *Farmer's Magazine* published a letter to Richardson from the Duke of Hamilton's factor, Robert Brown, about using fiorin to stabilise Hebridean 'blowing sands'. But Brown's letter was not printed until Richardson, worried that Brown might feel snubbed, had to utilise Young's influence to ensure that query and reply were published.[132]

Trapped between the need to solidify support in Scotland and to create it in England, Richardson resorted to the polemics of the electioneering squib, creating enemies without gaining friends. A Southampton gentleman, Thomas Hobbes Scott, had accompanied the Kirkcudbrightshire delegation to Clonfeacle. Richardson crowed in the *Agricultural Magazine* about how his visitors included gentlemen from all over the United Kingdom, but tried to drive his point home by manipulating national pride. Kirkcudbrightshire's 'patriotic endeavours' were received much better in Scotland than in England, where farmers were 'confident in their agricultural skills, and tenacious of old usages' but suspicious of 'novelties'. Scots were more confident in experimentation. Kirkcudbright's enthusiasm exposed Bath's hesitation. Pride made English farmers reluctant to admit that nature's gift lay overlooked and botanical confusion made them mistake fiorin for weeds. Sneering Englishmen, he wrote in this English periodical, should be 'shamed' by fiorin's success in the Celtic fringes. Fury at such Irish arrogance rather than shame characterised the ripostes, including that from the farmer who stated that Richardson's ideas would not work in 'south Britain' though they might satisfy 'a native bog-trotter'.[133]

Richardson's loss of control of knowledge was again exposed when the *Agricultural Magazine* carried lukewarm reports of fiorin experiments at Woburn by the Duke of Bedford's head gardener, George Sinclair. This infuriated Richardson, who complained that it had appeared without his permission and he vowed to 'thrash the hirelings' responsible.[134] Clearly he had not relinquished his original plan of accessing aristocratic opinion and obtaining parliamentary assistance. Yet by 1813, despite successes in Ireland, northern England, Wales and Scotland and interest overseas, overall recognition remained elusive. The ubiquity problem was a major contributing factor in fiorin's English failure. Criticisms in William Curtis's *Practical observations on British grasses* (1812) were endorsed by the botanist Dr Thornton that English fiorin was 'far inferior to Irish'.[135] Richardson was inexorably drawn into reacting to criticism rather than setting the agenda. Yearning for personal contact with elite patrons whose respectability would overawe doubters, he met an informed agricultural public opinion mediated through periodicals and associations. Something different was needed. In 1813 a major *New essay* became a watershed, but in an unwelcome way.

Fiorin grass: combating prejudice

The title was misleading. Familiar themes were rehashed. British 'aborigines' knew of fiorin, but modern writers like Arthur Young and Davies of Wiltshire discouraged people by giving derogatory names to the grass, which, if tried, would reveal its qualities. Linnaean botanists fixated with classification and nomenclature failed to recognise its properties. Parliament should promote fiorin improvement, as the public would willingly pay to enhance national strength and prosperity. Smarting from queries about fiorin's productivity, Richardson filled pages with statistics seeking to prove that fiorin would stop the financial drain of the poor laws. The wartime shortage of shipbuilding timber was grafted to the argument that fiorin and plantations could coexist. His triumphs, ranging from Shotts Iron Works to Hertford's mountain, were ludicrously compared to Henry IV defending his hereditary kingdom.[136]

Though Richardson was oblivious to this absurdity, readers were not; and, in a sudden switch of emphasis, the once-supportive *Farmer's Magazine* turned against him in a hostile review. Mocking Richardson's

Shakespearean analogy, like many 'great heroes' his irascibility could be excused by his battles in defence of his protégé. Anachronistic chauvinism was also ridiculed. Fiorin had advanced so slowly in Scotland that more acreage could be covered by the mass of 'essays, letters and dissertations' it generated than the grass itself. 'The honour of our country' was already satisfied by the assertion that Scots were more open-minded than their English neighbours. But, if fiorin was so wonderful, why did intelligent men oppose it? Turning to facts, the review noted that Scottish and English readers were rightly sceptical regarding Richardson's calculations of fiorin's potential contribution to national prosperity and claims about feeding cattle. Fiorin was unsuitable for arable land and feasible only in miry bog. Richardson knew little of modern husbandry, used the 'language of enthusiasm' and provoked opposition by his arrogance.[137]

This was a telling point. Despite his rebarbative bluster, Richardson was deeply worried about what appeared in periodicals and newspapers. Essentially fiorin had to compete on merit rather than on its socially constructed reputation. Responding to an Englishman's identification of new plants in the *Agricultural Magazine*, Richardson sourly wondered if he would meet the same ignorant prejudice as he had faced. In the same journal, one letter writer demanded that Richardson state exactly how much weight was lost in winter haymaking. Even the etymology of fiorin faced scrutiny, with the implication that it was contrived. Richardson backtracked and admitted that he did not understand the Irish language, and may have spelt fiorin incorrectly, as its phonetic rendition was 'fyorin'. Rather than being a platform for praise, county reports from the Board of Agriculture charted fiorin's decline. A Derbyshire report remarked that fiorin should not go from being overrated to underrated, as farmers found it growing locally but got meagre crops which cattle disliked.[138] Other supporters denied the perceived national differences, claiming that Humphry Davy's chemical analysis showed that English and Irish fiorin had a similar saccharine content.[139] But such pleas for open-mindedness marked how far opinion had turned against fiorin. Publishers sensed this, and Harding refused to buy out the copyright for a collection of essays.[140]

Thwarted in meadow and market, Richardson belatedly began to consider the arguments against him. Clutching at scientific straws instead of teaching improvers how to sow stolons, he retreated into pure experimentalism. He told the long-suffering Greenough that 'I have often complained … that in the school of agriculture, considered as a science, a personage is wanting, the experimentalist; [and] that this place was occupied by another, quite unequal to the task, the *Practical Farmer*'. He had laboured to revive experimentalism, and promised 'the world the benefit likely to be derived from the science of agriculture'. An early disciple, Lord Rous, who had experimented with fiorin on his Suffolk salt marsh, now intended to use it for livestock. On the classification question Richardson fell into the arms of his old enemies, the scientific botanists. Making a virtue of a necessity he even accepted Dr Thornton's opinion of differences between *Agrostis stolonifera* and *Agrostis alba*. Richardson claimed his own experiments had shown the latter to be useless, and hoped Thornton's scientific authority should 'crush these ignorant pretenders to botanical knowledge'. These were ideological somersaults; but Richardson ended up falling in front of those he had once courted. A 'Practical Farmer' mocked Richardson's despairing conversion, asking why anyone would listen to a 'chemist' when they had tangible results in their own soil. Richardson had used the same argument when, in his persona as the practical man of the land, he called Rennie a mere 'chemist' who 'studies in his laboratory alone'. A Suffolk farmer asked how Richardson had the audacity to impose 'trash' on Englishmen, useless, 'except in bogs too rotten to admit stock'. Regarding the plan for parliamentary money, 'having seen and experienced' fiorin, the same farmer thought it should 'remain at John O'Groat's house, where Dr. Richardson says it has extended to'. Another correspondent wondered how anyone could contradict 'such … respectable testimony'; yet, whatever the Irish position, fiorin was useless in 'South Britain', and enlightened farmers would hardly prefer prejudice over profit.[141]

Richardson responded by sending samples of 'true fiorin' to Vauxhall to show how it could flourish alongside plantations of fruit trees. Another tactic was his determination to publish an 'annual statement' to counteract the 'obstinate incredulity' which 'deprived' England of an agent which would 'spread luxurious crops over its now barren and desolate tracts'.[142] Yet these initiatives received only a few lines in the press and, by 1814, his vision of improvement fading, Richardson resorted to grafting fiorin culture on to contemporary notions of manliness, even though the botanical implications contradicted his recent conversion to 'agriculture as a science'. Those who quibbled about varieties of fiorin suffered from 'Linnaean folly'; but true 'pupils', like his grass, were a hardy northern breed, uninterested in Linnaeus's system of sexual classification. In contrast to his critics' prurience, supporters did not 'know what a panicle was', nor had 'the least curiosity about how grasses propagate'. What these hardy northern farmers did know was how to plant fiorin effectively. *Farmer's Magazine* readers should consider that 'in the North of Ireland we never think of *varieties*, but always hit the *right* sort'. Timorous English agriculturalists joined 'the labouring Linnaeans' to discover no less than forty-three *Agrostis* varieties but they should not 'infect the south of Ireland with their wisdom'. Lustiness was acceptable but not lasciviousness. Fiorin was a 'true-spirited soldier' who endures 'the fatigues of a campaign and enjoys the luxuries of a city'. This appeal to manliness underlay his comment that 'when we [the fiorin cultivators] send out our foraging parties we never remind the old ladies to take their spectacles with them'. However, this approach glorified anti-intellectualism and backfired badly in enlightened Scotland, where 'A. Clodhopper' complained that Richardson was elevating 'the pre-eminence of ignorance over knowledge'.[143]

Even solid Scottish supporters like Aiton of Strathaven grew uneasy, embarrassed by Richardson's rhetorical fisticuffs, which he tried to excuse as a reaction to 'unreasonable opposition'. Even he conceded that Richardson had exaggerated fiorin's ability to thrive everywhere, even if it outperformed other crops on marshy grounds.[144] But statistics were closing in on rhetoric. A Board of Agriculture report from Sinclair's Caithness noted that though fiorin grew spontaneously, 'how far it will discover in this county, equal to what has been said regarding its produce in Ireland remains to be ascertained'.[145] Vulnerable to such critiques, Richardson insisted that people follow his instructions minutely about correct cultivation, otherwise he disavowed the outcome. Alexander Young of Edinburgh mowed too early, exasperating Richardson, who wanted him to wait till October. Even converts like Curwen backtracked, claiming that fiorin worked only on peat bogs.[146] As criticisms mounted in northern Britain, matters took a turn for the worse in Ireland.

Sir John Sinclair had once quoted Bacon's dictum that 'knowledge is power'.[147] But, with control over fiorin knowledge gone, Richardson was powerless to prevent adverse information spreading back to Ireland. Complaints about livestock rejecting fiorin hay reached Abercorn's land steward, who reported that cattle refused it and those which did not produced 'scarce a drop of milk'. Irrigation was ineffective. Where the subsoil was retentive, water 'wore out' fiorin and necessitated expensive drains.[148] Though Richardson had boasted about 'how I fight the Cork men', bad English publicity flowing back along the periodical network increased opposition in Ireland.[149] His old adversary, Townsend, cited the *Agricultural Magazine*'s claim that Richardson exaggerated fiorin's productivity. Now complacently confident, Townsend indulged himself in a conceit, archly comparing himself and Richardson to fellow travellers in a paper conveyance having a polite disagreement. Affecting to concede in this gentlemanly joust, but actually mocking Richardson's pomposity, Townsend switched metaphor. Richardson became a daring acrobat who 'raises the astonishment of the spectators', while he was the clown who did 'little more than make them laugh'. But his real point was that before believing 'northern testimony' a 'southern man wanted the reports of southern experimentalists'. The classification problem had spread to Ireland. The readers of *Munster Farmer's Magazine* wondered if there was a distinct northern variety.[150] The Farming Society of Ireland's offer of a fiorin premium in 1815 was considered beneficial if it could do away with any 'delusion' over Richardson's claims, thus implying that cultivating the grass risked ridicule by association.[151]

Humiliated, but not defeated, in July 1815 Richardson turned to an ageing warhorse of aristocratic improvement, the gout-stricken Sir Joseph Banks. Fiorin was established in Ireland and Scotland, but not in England, 'where an inferior variety of *Agrostis* has discouraged many'. The failure with English opinion was undeniable. The Bath and West of England Society's experiment had failed miserably, its premium winner producing only 'foul' grass, and Richardson's own irrigation claims were disproved. Trying to salvage any utility, he argued that good whiskey could be distilled from fiorin; but Banks sensibly advised an exit strategy: Richardson should claim all *Agrostis* species; then withdraw, content with having been 'being a benefactor to mankind'.[152] Retreat, however, was the last thing on his mind. In return for his advice, Richardson sent Banks an essay on fiorin in Newfoundland.

This had been solicited by the colony's Governor, Sir Richard Keats, and its garrison commander, Lieutenant-General Moore. Richardson was 'gratified' by a request from so valuable a colony, where he assumed fiorin grew naturally and promised better hay than any in England. Attempting to regain control and perhaps assuming that the traditional elite role in improvement survived in the colonies, Richardson rejoiced that the request came from military men who could obey orders. Newfoundland amateurs should follow his directions and ignore bad impressions from England attributable to the agriculturists, not the grass.[153] This venture may have appealed to Banks's imperialism, but Richardson ignored his hints to desist at home. Richardson had trumpeted Danish interest in fiorin, but a Hampshire farmer said such nonsense would dissuade people from testing whether 'the Irish grass' had minimal utility as winter food. 'J.B.' of Sussex noted that, while fiorin might work in bogs, 'as to encumbering our best grass lands with English or Irish fiorin, the good lord deliver us from both English and Irish insanity'; academic points might be discussed but, essentially, 'the business is at rest'. Mockery followed marginalisation. In 1815 one of the *Agricultural Magazine*'s 'scribblers', S.T. of Norfolk, lampooned him mercilessly in a mock-heroic style: 'O Richardson! Whose matchless deeds,/To every land and clime extend,/Tell us, immortal man of weeds,/Where will thy chain of wonders end?'[154]

Though Richardson's star declined in Britain, he retained hopes overseas. The Copenhagen businessman and trader Frederick de Conynck, to whom Richardson had written in 1811, engaged in a correspondence about cultivation methods and published the results of successful experiments and his exchanges with Richardson. 'This important grass' would 'transform Danish agriculture' and he introduced Johannes Christian Drewsen, a Swedish paper manufacturer and agriculturalist, to fiorin. Drewsen was impressed by the quantities of hay de Conynck had achieved. His initial trial failed, but he too wrote to Richardson for advice about ploughing and weeding. A second attempt succeeded well, prompting Drewsen to dedicate a pamphlet on 'this remarkable grass' to the Swedish minister of state for agriculture. Swedish agricultural societies were interested and Drewsen believed that fiorin was 'the most important plant that the nineteenth century has taught us to use'.[155] This European publicity led to an unexpected and inviting offer in November 1817.

The Archduke of Austria had asked de Conynck for samples and, when visiting London, wanted the MP Sir Thomas Ackland to get Irish fiorin for comparison.[156] Though Richardson was considering ending his agricultural 'exploits', neither he nor his publisher, Whitmore & Fenn, could resist the Archducal imprimatur. In a lengthy compilation Richardson noted the flattering attention of 'our imperial ally', thanked 'pupils' in Denmark, Sweden and Norway, and noted how their experiments generated enquiries from New York and Boston. He compiled his knowledge on fiorin's qualities and uses, and noted how opposition from the 'practical farmer' and his instructors the 'seedsmen, nurserymen, and agricultural book-makers, mostly from GRUB STREET' was invalidated by the testimony of scientists, noblemen, bishops, generals, MPs and agricultural societies. In Richardson's eyes this work may have crowned his aristocratically orientated vision of improvement, but the *Farmer's Magazine* consigned it to 'the annals of empiricism – or lunacy'.[157]

Fiorin's failure?

This kind of dismissal continued after Richardson's death. John Donaldson's *Agricultural biography* (1854) called his work ephemeral and careless.[158] Yet such portrayals were profoundly anachronistic, arising from professionalised science's disparaging backward glance at an eighteenth-century 'amateur'. The geographical range of fiorin knowledge alone gives the lie to Donaldson but, more importantly, such verdicts ignore the contemporary contexts in which fiorin knowledge was produced, focusing on the outcome rather than the twelve-year battle for opinion. As even its critics admitted, Richardson's advocacy created a national controversy. To fully understand this we must view the campaign in its broadest cultural context.

The key point is that Richardson's fiorin campaign occurred in the immediate aftermath of the Union. Historians recognise the Union's political impact in making issues like Catholic emancipation more prominent on the government's agenda.[159] However, its intellectual implications tend to be overlooked, being less visible. This reconstruction of Richardson's campaign shows that the Union presented inviting opportunities for Irishmen. He deliberately used the influx of Irish MPs and peers to Westminster to channel improving knowledge straight to the centre. The fact that Abercorn, Downshire and Hertford were quintessentially Anglo-Irish also helped. Their social and political relevance, plus the privilege of parliamentary franking, made them ideal transmitters of knowledge.

Despite relative success in Scotland, with its congenial natural and cultural environment, Richardson never overcame the problem of reconstructing Irish knowledge for a British audience. He addressed the logistical difficulties of isolation through contacts in London and Edinburgh, but was drawn into a more modern and commercialised setting of periodical publication and agricultural associationalism. He entered these spheres reluctantly and performed indifferently, hankering for unmediated access to elite improvers. When thwarted, he reverted to socially constructed interpretations, blaming mercenary booksellers and the Grub Street 'scribblers'. Though he courted newer forms of knowledge community and sought to enmesh himself in their networks, he missed both the personal contact of elite patronage and not having to navigate associationalism or negotiate with public opinion. Richardson told Isaac Corry of the difference when 'I chat to a familiar friend' and the 'stiff erect posture, in which I address the grave Dons of literary and agricultural societies'.[160]

Though intervention in periodicals seemed to suit Richardson's combative nature, as time went on an inverse relationship developed between the wide coverage he wanted and his control over knowledge. When contradicted he reverted to a polemical style which subverted the economic patriot's logical ambitions to chauvinistic national patriotism, attempting to force compliance by embarrassment. But though this

Protestant Irishman yielded to no one, he badly miscalculated attitudes in Britain and provoked equally prejudiced reactions. Richardson assumed that Scots, a century after their Union, were as jealous of their identity as Irishmen were a decade after theirs, and ignored the Scottish Enlightenment's progressive cosmopolitanism and the development of a 'north British' identity predicated on political and economic assimilation with England.[161]

Arrogant rhetoric diverted attention from fiorin and excess zeal led to charges of utopianism. Richardson may have even unconsciously contributed to the failure, as several critics pleaded that his exaggerations should not disguise fiorin's limited use in specific contexts. It is tempting to see north–south cultural differences, even in Ireland, with the Scottish influence in Ulster and the more Anglicised Munster. However, environmental determinants were more important. Every area of relative success – Scotland, Ulster, Wicklow, Cumberland, Dartmoor, Wales and the Isle of Man – had extensive terrain deemed capable of improvement. During the wartime agricultural crisis Richardson offered a remedy which was taken very seriously by influential people and projected himself and his grass to the notice of national opinion. Agriculture was not the only contemporary area of concern. Malthus's essay on population turned many minds, including Richardson's, towards political economy.

Notes

1 *Agricultural Magazine*, new series, vol. 5 (January–June 1815), p. 178.
2 *Farmer's Magazine*, vol. 75 (August 1818), pp. 346–8.
3 A. Sneddon, 'Legislating for economic development: Irish fisheries as a case study of the limits of "improvement"', in D. W. Hayton, J. Kelly and J. Bergin (eds), *The eighteenth-century composite state: representative institutions in Ireland and Europe, 1689–1800* (Basingstoke, 2010), p. 137.
4 C. W. J. Withers, 'Improvement and enlightenment: agriculture and natural history in the work of the Rev. Dr. John Walker (1731–1803)', in P. Jones (ed.), *Philosophy and science in the Scottish Enlightenment* (Edinburgh, 1988), p. 102.
5 Richardson, *Letter to the Rt. Hon. Isaac Corry*.
6 P. Borsay, 'The culture of improvement', in P. Langford (ed.), *The eighteenth century* (Oxford, 2002), pp. 185–7; Withers, 'Improvement and enlightenment', p. 106.
7 Golinski, *Science as public culture*, pp. 32–5.
8 Borsay, 'Culture of improvement', p. 191; D. P. Miller, 'The usefulness of natural philosophy: the Royal Society and the culture of practical utility in the later eighteenth century', *BJHS*, vol. 32 (1999), pp. 185–201.
9 R. L. Emerson, 'The Philosophical Society of Edinburgh, 1768–83', *BJHS*, vol. 18 (1985), p. 270.

10 R. Mitchison, *Agricultural Sir John: the life of Sir John Sinclair of Ulbster* (London, 1962), pp. 135, 141–2; R. Mitchinson, 'The old Board of Agriculture, 1793–1822', *English Historical Review*, vol. 74 (1959), p. 44; B. Hilton, *A mad, bad and dangerous people? England, 1783–1846* (Oxford, 2006), p. 135.

11 Borsay, 'Culture of improvement', pp. 197–8.

12 Withers, 'Improvement and enlightenment', pp. 102, 106.

13 Mitchison, *Agricultural Sir John*, p. 103; Withers, 'Improvement and enlightenment', pp. 108, 111–13; Emerson, 'Philosophical Society of Edinburgh', pp. 258, 293.

14 N. Phillipson, *Adam Smith, an enlightened life* (London, 2010), pp. 39–40, 128–30; Mitchison, *Agricultural Sir John*, p. 23; Withers, 'Improvement and enlightenment', p. 102; Emerson, 'Philosophical Society of Edinburgh', p. 271.

15 Michael S. Moss, 'Miller, Patrick (1731–1815)', *Oxford dictionary of national biography* (Oxford, 2004), online edition, October 2007, www.oxforddnb.com/view/article/18733, accessed 12 November 2010.

16 Malcomson, *John Foster*, pp. 378–80, 373.

17 E. Magennis, 'A land of milk and honey: the Physico-Historical Society, improvement and the surveys of mid-eighteenth-century Ireland', *Proceedings of the Royal Irish Academy*, vol. 102C, no. 6 (2002), pp. 199–217; T. Barnard, 'The Dublin Society and other improving societies, 1731–85', in J. Kelly and M. J. Powell (eds), *Clubs and societies in eighteenth-century Ireland* (Dublin, 2010), pp. 69–70.

18 Minutes of the Dublin Society, 22 January 1801, 19 March 1801, Royal Dublin Society Library.

19 P. Walsh, 'Club life in late seventeenth-century and early eighteenth-century Ireland', in Kelly and Powell (eds), *Clubs and societies*, p. 41; D. A. Fleming, 'Diversions among the people: sociability among the orders of early eighteenth-century Ireland', *ECI*, vol. 17 (2002), pp. 99–111.

20 Barnard, 'The Dublin Society', p. 68; K. Milne, 'Irish charter schools', *Irish Journal of Education*, vol. 7 (1974), pp. 3–29.

21 M. Brown, 'Configuring the Irish Enlightenment: reading the *Transactions of the Royal Irish Academy*', in Kelly and Powell (eds) *Clubs and societies*, pp. 168, 178; C. Maxwell, *Dublin under the Georges* (London, 1956), p. 213.

22 Malcomson, *John Foster*, p. 375.

23 Barnard, 'The Dublin Society', pp. 58–64, 82; J. Livesey, 'The Dublin Society in eighteenth-century Irish political thought', *HJ*, vol. 47 (2004), pp. 615–40; E. Magennis, 'Coal, corn and canals: parliament and the dispersal of public moneys, 1695–1772', in D. Hayton (ed.), *The Irish parliament in the eighteenth-century* (Edinburgh, 2000), p. 78; J. McEvoy, *Statistical survey of the county of Tyrone* (Dublin, 1802); K. Bright, *The Royal Dublin Society, 1815–1845* (Dublin, 2004), pp. 15, 27, 30–3; T. de Vere White, *The story of the Royal Dublin Society* (Tralee, 1955); Malcomson, *John Foster*, p. 375.

24 Magennis, 'Coal, corn and canals', pp. 71–86; Sneddon, 'Legislating for economic development', pp. 127–8; T. Barnard, *The kingdom of Ireland, 1641–1760* (Basingstoke, 2004), p. 85.

25 Magennis, 'Coal, corn and canals', p. 86; Bright, *Royal Dublin Society*, pp. 15, 27.

26 E. A. Smith, 'Russell, Francis, fifth Duke of Bedford (1765–1802), agriculturist and politician', *Oxford dictionary of national biography* (Oxford, 2004), online edition, www.oxforddnb.com/view/article/24308, accessed 12 July 2011.

27 Emerson, 'Philosophical Society of Edinburgh', pp. 7, 292; Borsay, 'Culture

of improvement', pp. 198–9; C. Kidd, 'North Britishness and the nature of eighteenth-century patriotisms', *HJ*, vol. 39, no. 2 (1996), pp. 362, 364, 375.

28 Mitchison, *Agricultural Sir John*, pp. 6, 90; Borsay, 'Culture of improvement', pp. 205–6; Richardson to Greenough, 17 December 1812, UCL, Additional Greenough papers, 1474.

29 D. Dickson, *Old world colony: Cork and south Munster, 1630–1830* (Cork, 2005), pp. 173, 176, cited in R. B. McDowell, *Ireland in the age of imperialism and revolution, 1760–1801* (Oxford, 1979), pp. 6–7.

30 P. Kelly, 'The politics of political economy in mid-eighteenth-century Ireland', in S. J. Connolly (ed.), *Political ideas in eighteenth-century Ireland* (Dublin, 2000), p. 113.

31 Sneddon, 'Legislating for economic development', pp. 137–8; A. Sneddon, 'Bishop Francis Hutchinson (1660–1739): a case study in the culture of eighteenth-century improvement', *IHS*, vol. 35, no. 139 (May 2007), pp. 300–6, 309; J. Hoppit, 'The contexts and contours of British economic literature', *HJ*, vol. 49, no. 1 (2006), p. 106.

32 T. Barnard, 'Improving clergymen, 1660–1760', in A. Ford, J. McGuire and K. Milne (eds), *As by law established the Church of Ireland since the Reformation* (Dublin, 1995), pp. 136–7, 146; T. Barnard, *Improving Ireland: projects, prophets and profiteers, 1641–1786* (Dublin, 2008), p. 13.

33 Sneddon, 'Bishop Francis Hutchinson', pp. 289–310; Connolly, *Religion, law and power*, pp. 299–300; Malcomson, *John Foster*, p. 354.

34 P. J. Jupp, *British politics on the eve of reform* (Basingstoke, 1998), pp. 331–5.

35 C. H. Weiss, 'Reflections on nineteenth-century experience of knowledge diffusion', *Science Communication*, vol. 13, no. 1 (September 1991), pp. 3–8.

36 Frissell and Cooter, 'Exploring natural knowledge', pp. 131–4.

37 Royal Dublin Society database, www.rds.ie/cat_historic_members.jspDatabase, accessed 26 August 2011.

38 McEvoy, *Statistical survey of the county of Tyrone*, pp. 13, 56.

39 W. Richardson, *An elementary treatise on the indigenous grasses of Ireland, with a selection of those which promise to be the most useful. Addressed to his agricultural friends* (Dublin, 1806), pp. 1, 6–9, 31–2.

40 Malcomson, *John Foster*, pp. 348, 373.

41 Bright, *Royal Dublin Society*, pp. 15, 27.

42 Minutes of the Dublin Society, 22 November 1798, 7 May 1801, Royal Dublin Society Library.

43 Hilton, *A mad, bad and dangerous people?*, p. 135; W. Richardson, *Plan for reclaiming the Bog of Allen and the other great morasses in Ireland addressed to the Right Hon. Earl of Rosse* (Dublin, 1809), pp. 3–4.

44 I have been unable to locate the 1807 Board of Agriculture memoir, but Richardson referred to it frequently elsewhere: *An essay on the improvement of the great flow bogs of Ireland particularly the Bog of Allen and the Montiaghs of the north … in a letter addressed to the Grand Juries of Antrim, Armagh and Tyrone* (Dublin, 1807); *Plan for reclaiming the Bog of Allen*; *Farmer's Magazine* (December 1809), p. 503; Richardson to Abercorn, 3 May 1808, PRONI, Abercorn papers, D623/A/165/29. He also published a supplement: *Comment and notes on the preceding memoir as published by the Board of Agriculture* (Coleraine, 1808).

45 Richardson, *Essay on the improvement of the great flow bogs*, pp. 6–7, 9, 11, 17, 19; Richardson, *Plan for reclaiming the Bog of Allen*, p. 4.

46 Richardson, *Memoir on fiorin grass*, pp. 5–12, 14–15, 20–25, 29–31; J. Archer, *Statistical survey of the county of Dublin* (Dublin, 1801), p. 131.
47 Barnard, 'The Dublin Society', pp. 62, 80.
48 Richardson to Arthur Young, 14 March 1809, BL, Arthur Young papers, add mss 35130, fol. 184.
49 Malcomson, *John Foster*, p. 358.
50 Banks to Young, 10 August 1809, Banks to Richardson, 14 August 1809, in Chambers (ed.), *Scientific correspondence of Sir Joseph Banks*, vol. 6, pp. 12–13.
51 W. Wade, *Sketch of lectures on meadow and pasture grasses* (Dublin, 1808), p. 45; J. Smith, *Essay on indigenous grasses* (Dublin, 1808), pp. 26–7; E. Wakefield, *Account of Ireland statistical and political* (London, 1812), vol. 1, pp. 462–3; Bright, *Royal Dublin Society*, p. 42.
52 Malcomson, 'A lost natural leader', p. 68; A. P. W. Malcomson, 'Introduction to the Abercorn papers in PRONI', pp. 7–8, 15–18, PRONI, Abercorn papers, D623.
53 Richardson to Abercorn, 3, 18 May 1808, PRONI, Abercorn papers, D623/A/165/29, 33/232.
54 Burgoyne to Abercorn, 14, 18 June, 3 July 1808, Richardson to Abercorn, 21 June, 3, 4 July 1808, PRONI, Abercorn papers, D623/A/123/39, 40, 43, D623/A/165/33–4.
55 Richardson, *Comment and notes on the preceding memoir*.
56 Richardson to Abercorn, 4, 10, 16 July 1808, Burgoyne to Abercorn, 1 March 1809, PRONI, Abercorn papers, D623/A/165/35, 36, 38, 8, D623/A/112/12.
57 Richardson to Joy, 23 May 1808, 17 October 1808, Joy to Richardson, 5 October 1808, LHL, Joy MSS, ix; Richardson to Abercorn, 22 October 1808, PRONI, Abercorn papers, D623/A/165/44.
58 Banks to Smith, 13 January 1809, in W. R. Dawson, *The Banks letters* (London, 1958), p. 766.
59 *Farmer's Magazine*, vol. 9 (1808), p. 340; Sir John Sinclair, *Address to the Board of Agriculture* (London, 1808), p. 4.
60 Richardson to Aberdeen, 20 November 1809, BL, Aberdeen papers, add mss 43230, fols 50–1; Mitchison, 'The old Board of Agriculture', p. 56.
61 Richardson to Greenough, 27 October 1809, UCL, Additional Greenough papers, 1428.
62 Malcomson, *John Foster*, pp. 291, 309, 323.
63 Richardson, *Plan for reclaiming the Bog of Allen*, pp. 3–4.
64 Richardson to Greenough, 18 May 1809, UCL, Additional Greenough papers, 1424; *Cobbett's parliamentary debates, first series* (1809), vol. 14, pp. 337–8.
65 Richardson to Greenough, 18 May, 12 June 1809, UCL, Additional Greenough papers, 1424, 1425; W. Richardson, *Essay on the culture of salt marshes* (Belfast, 1810), p. 34.
66 Richardson to Abercorn, 4, 10 July 1808, PRONI, Abercorn papers, D623/A/165/35, 36; Richardson to Aberdeen, 13, 18 September 1808, BL, Aberdeen papers, add mss 43229, fol. 333, 43230, fol. 17; Aberdeen to Abercorn, 16 November 1808, PRONI, Abercorn papers, T3472/2/11.
67 Richardson to Aberdeen, 6, 10 October 1809, BL, add mss 43230, fols 41–2, 43–4.
68 Richardson to Aberdeen, 6, 10 October 1809, BL, Aberdeen papers, add mss 43230, fols 41–2, 43–4; A. Edmundston, *A view of the ancient and present state of the Zetland Islands* (Edinburgh, 1809), p. 360.
69 Richardson to Young, 1 March 1809, NAS, Hannay papers, GD214/736–7.

70 Michael S. Moss, 'Miller, Patrick (1731–1815)', *Oxford dictionary of national biography* (Oxford, 2004), online edition, October 2007, www.oxforddnb.com/view/article/18733, accessed 12 November 2010; P. Miller (Senior) to P. Miller (Junior), 22 March 1808, 3, 13 September 1809, NAS, Miller of Dalswinton papers, GD197/4/2/2/28, 29, 308; J. Farish, *Treatise on fiorin grass with a short description of its nature and properties, together with the soils and manures, best adapted to its culture* (Dumfries, 1810), p. 5.

71 W. Richardson, *Essay on the improvement of Dartmoor Forest, and on the means of so much increasing our grain crops as to make future importation unnecessary; both to be attained by the aid of fiorin grass*, printed in *Letters and papers on agricultural planting selected from the correspondences of the Bath and West of England Agricultural Society* (Bath and London, 1814), vol. 13, p. 156.

72 Richardson to Aberdeen, 20 November 1809, BL, Aberdeen papers, add mss 43230, fols 50–1.

73 Jupp, *British politics*, pp. 331–5.

74 *Farmer's Magazine*, vol. 10 (1809), pp. 503–9.

75 *Agricultural Magazine*, vol. 4 (1809), pp. 306–7.

76 Richardson to Aberdeen, 20 November 1809, BL, Aberdeen papers, add mss 43230, fols 50–1.

77 *Farmer's Magazine*, vol. 41 (1810), pp. 183–4.

78 J. Hoppit, 'The context and contours of British economic literature, 1660–1760', *HJ*, vol. 49, no. 1 (2006), p. 96.

79 Richardson to Greenough, 10 October 1810, 15, 22 November 1809, UCL, Additional Greenough papers, 1449, 1431, 1430; W. Richardson, *Letter on irrigation to the Rt. Hon. Isaac Corry* (Belfast, 1810).

80 Richardson to Greenough, 22 November 1809, UCL, Additional Greenough papers, 1431; Richardson to Aberdeen, 20 November 1809, BL, Aberdeen papers, add mss 43230, fols 50–2; W. Richardson, *Epitome of a letter on irrigation addressed to the Rt. Hon Isaac Corry* (London, 1810).

81 W. Richardson, *Letter on the subject of reclaiming and improving the waste parts of the Highlands of Scotland … by introducing the culture of fiorin grass* (London, 1810); Richardson to Greenough, 26 December 1809, 4 February, 1 June 1810, UCL, Additional Greenough papers, 1432, 1436, 1443; Richardson to Aberdeen, 14, 17 December 1809, BL, Aberdeen papers, add mss 42330, fols 62, 64–5.

82 Richardson to Greenough, 26 December 1809, 4 February 1810, UCL, Additional Greenough papers, 1432, 1435; Richardson to Aberdeen, 9 October 1809, BL, Aberdeen papers, add mss 42330, fols 60–1; *Agricultural Magazine*, vol. 11 (1810), p. 379.

83 Sneddon, 'Legislating for economic development', pp. 137–8.

84 Minutes of the Dublin Society, 30 November 1809; Royal Dublin Society database, www.rds.ie/cat_historic_member_detail.jsp?itemID=1096610&item_name=Corry, accessed 19 March 2012.

85 M. D. George, *Catalogue of political and personal satires* (London, 1952), p. 884.

86 Richardson to Joy, n.d. [c. October 1809], LHL, Joy MSS; Malcomson, *John Foster*, p. 88.

87 Richardson, *Letter on irrigation to the Rt. Hon. Isaac Corry*, pp. 1–47.

88 Richardson, *Essay on the culture of salt marshes*, pp. 1–18, 23–24; Richardson to Greenough, 22 November 1809, 15 March 1810, UCL, Additional Greenough papers, 1431, 1437; Mitchison, *Agricultural Sir John*, p. 208.

89 Richardson to Downshire, 22 February, 8 March 1810, PRONI, Downshire papers, D671/C/335/1A, 2.

90 Richardson to Greenough, 5 July, 9 October, 14 November 1810, UCL, Additional Greenough papers, 1445, 1448, 1450–1; W. Richardson, *Letter to the most noble the Marquis of Hertford on fiorin grass containing the necessary directions for its culture; the periods and modes of laying it down....* (London, 1810), pp. 3–7, 15, 20, 31, 43.

91 *Belfast News-Letter*, 30 January, 9 February 1810; Richardson to Greenough, 4 February, 1 April 1810, UCL, Additional Greenough papers, 1435, 1437.

92 Farish, *Treatise on fiorin grass*, pp. 1, 3, 6–7, 11, 18, 27, 33.

93 Richardson to Greenough, 4 February 15 March, 20 April, 1 June, 21 August, 9, 10 October 1810, UCL, Additional Greenough papers, 1436, 1437, 1441, 1443, 1446, 1448, 1449 .

94 Richardson to Greenough, 15 November 1809, UCL, Additional Greenough papers, 1430.

95 Richardson to Nicholl, Richardson to Greenough, 21, 29 August 1810, UCL, Additional Greenough papers, 1446, 1447.

96 Richardson to Greenough, 5 February 1811, UCL, Additional Greenough papers, 1453.

97 *Farmer's Magazine*, vol. 12 (June–August 1811), pp. 227–30, 354–9.

98 *Lancaster Gazette*, 12 October 1811, 14, 28 July 1812.

99 Richardson to Kater, 18 January 1813, Suffolk County Record Office, HA231/3/1/114.

100 Minutes of the Middleward of Lanarkshire Farmers' Club, 27 April 1807, NAS, Douglas-Hamilton papers, NRAS2177/944.

101 *Caledonian Mercury*, 7 November 1812.

102 R. Ainslie, *Memoir on the fiorin grass* (Kirkcudbright, 1811).

103 *Lancaster Gazette*, 12 October 1811.

104 Richardson to Greenough, 11 November 1811, UCL, Additional Greenough papers, 1471; Richardson, *Essay on the improvement of Dartmoor*, p. 130.

105 Moss, 'Miller, Patrick (1731–1815)'; *Caledonian Mercury*, 6 January 1812.

106 *Rules and proceedings of anniversary of the Workington Agricultural Society* (Workington, 1811), pp. 64, 161, 184–6, 203–26; Richardson to Greenough, 3 November 1811, UCL, Additional Greenough papers, 1470; *Lancaster Gazette*, 12 October 1811.

107 Richardson to Greenough, 7 January, 3 November 1811, UCL, Additional Greenough papers, 1452, 1470.

108 *Farmer's Magazine*, vol. 49 (1812), pp. 1–2, 3–25.

109 *Norfolk, Essex, Cambridge and Ely Advertiser*, 12 February 1812.

110 *Agricultural Magazine*, vol. 8 (January–June 1811), pp. 18, 49, 71, 125, 262, 320.

111 *Weekly Entertainer or Agreeable and Instructive Repository*, vol. 52, (1812), pp. 810–13; E. Bronson (ed.), *Select Reviews of Literature*, vol. 7 (1812), pp. 64–5.

112 Jefferson, cited in John Chambers to Robert Simms, 24 May 1811, PRONI, Emigrant letters, T1815/12A.

113 *Belfast News-Letter*, 18 June 1811; Richardson to Greenough, 15 May 1811, UCL, Additional Greenough papers, 1458.

114 Richardson to Nichols, Richardson to Pictet, 24 April 1811, Richardson to Greenough, 15 May 1811, UCL, Additional Greenough papers, 1455, 1456, 1458.

115 Sheffield to Foster, 7 May 1803, Sir John Sinclair to Foster, 5 February 1808, PRONI, Foster–Massereene papers, D562/7891, D562/5834.
116 *Freeman's Journal*, 12 September 1811; W. Richardson, *The utility of fiorin grass: a prize essay addressed to the Farming Society of Ireland, on their proposing to give a gold medal 'to the person who shall state to the Society the result of an experiment which shall be the most satisfactory in ascertaining the merit and utility of fiorin grass etc'* (London, 1811); *Morning Chronicle*, 20 March 2012.
117 Richardson, *Utility of fiorin grass*, pp. 1–9; 13–14, 20–3, 25–31, 34; Richardson to Greenough, 6 February 1813, UCL, Additional Greenough papers, 1469.
118 Malcomson, *John Foster*, p. 344.
119 W. Richardson, *Observations on a report of the agriculture and livestock of the county of Wicklow* (Dublin, 1812), pp. 8–10, 12–18, 21, 26–30, 32–4, 38, 43–6.
120 Dickson, *Old world colony*, pp. 289–90.
121 *Munster Farmer's Magazine*, vol. 1 (January–June 1811), pp. 31–2, 139–48, 213–23, 240, 333; Dickson, *Old world colony*, pp. 283, 517.
122 Wakefield, *Account of Ireland*, vol. 1, pp. 462–3.
123 *Munster Farmer's Magazine*, vol. 1 (January–June 1811), pp. 31–2, 139–48, 213–23, 240, 333; Dickson, *Old world colony*, pp. 283, 517.
124 Richardson to Greenough, 11 February 1811, 29 December 1812, UCL, Additional Greenough papers, 1454, 1476; Richardson, *Essay on the improvement of Dartmoor* (*passim*).
125 J. G. Lockhart, *The memoirs of Sir Walter Scott* (Paris, 1838), vol. 3, p. 141; W. Richardson, *Memoir on the cultivation of fiorin grass in a letter addressed to the Bath and West of England Agricultural Society* (Bath, 1812), pp. 1, 6–7, 9, 14, 19, 25–9.
126 *Belfast News-Letter*, 26 October 1810.
127 Richardson, *Memoir to the Bath and West of England Agricultural Society*, pp. 42–55.
128 Richardson, *The correspondence between the Rev. Dr. Richardson and Mr. Tanner on fiorin grass* (London, 1812), pp. 6–8, 14–18, 23–5, 31.
129 *Farmer's Magazine*, vol. 49 (1812), pp. 182–5, 201–9.
130 *Farmer's Magazine*, vol. 14 (1813), pp. 193–4; Richardson, *Essay on the improvement of Dartmoor*, pp. 133–6.
131 Richardson to Young, 25 January 1813, NAS, Hannay papers, GD214/736/8.
132 *Farmer's Magazine*, vol. 49 (1812), pp. 365, 467; Alexander Young to Robert Brown, 23, 25 November 1812, John Sheriff to Robert Brown, 1 July 1812, NAS, Douglas–Hamilton papers, NRAS2177/1584, 1575.
133 *Agricultural Magazine*, vol. 10 (January–June 1812), pp. 6–10, 133, 149–50.
134 *Agricultural Magazine*, new series, vol. 1 (January–June 1813), p. 16; Richardson to Greenough, 6 February 1813, UCL, Additional Greenough papers, 1479.
135 *Lancaster Gazette*, 24 April 1813.
136 W. Richardson, *A new essay on fiorin grass* (London, 1814), *passim*.
137 *Farmer's Magazine*, vol. 14 (1813), pp. 471–81.
138 *Agricultural Magazine*, new series, vol. 1 (January–June 1813), pp. 97–101, 309.
139 *Lancaster Gazette*, 17 July 1813.
140 Richardson to Greenough, 25 March 1813, UCL, Additional Greenough papers, 1484.
141 Dubordieu (ed.), *Statistical survey*, vol. 2, pp. 11, 101, 476; *Agricultural Magazine*, new series, vol. 1 (January–June 1813), pp. 313–15, 361–3, 401–3; Farish, *Treatise on fiorin grass*, p. 3.

142 *Morning Post*, 18 September 1813; *Derby Mercury*, 10 November 1814.
143 Richardson, *Essay on the improvement of Dartmoor*, p. 160; *Farmer's Magazine*, vol. 58 (1814), p. 192.
144 *Farmer's Magazine*, vol. 64 (1815), pp. 53–4.
145 J. Henderson, *A general view of the agriculture in the county of Caithness* (London, 1815), pp. 148–9.
146 Richardson to Young, 11 September 1815, NAS, Hannay papers, GD214/736/9/1–2; *Lancaster Gazette*, 10 June 1815.
147 Sinclair to Foster, 19 January 1813, PRONI, Foster–Massereene papers, D207/13/11.
148 Burgoyne to Abercorn, 11 September, 2 December 1813, 14 March 1814, 20 February 1815, PRONI, Abercorn papers, D623/A/126/4, 17, D623/A/127/2, D623/A/128/9; Richardson to Banks, 14 July 1815, in Chambers (ed.), *Scientific correspondence of Sir Joseph Banks*, vol. 6, pp. 176–7.
149 Richardson to Greenough, 17 January 1812, UCL, Additional Greenough papers, 1474.
150 *Munster Farmer's Magazine* (1813), pp. 11–15.
151 *Munster Farmer's Magazine* (1815), p. 232.
152 Richardson to Banks, 14 July 1815, in Chamers (ed.), *Scientific correspondence of Sir Joseph Banks*, vol. 6, pp. 176–7; *Belfast News-Letter*, 25 August 1815.
153 W. Richardson, *Directions for the cultivation of fiorin grass in the island of Newfoundland* (Newry, 1815), pp. 2–5, 13, 24.
154 *Agricultural Magazine*, new series, vol. 5 (January–June 1815), pp. 75–82, 111, 171–8.
155 F. de Conynck, *Fiorin Gräs eller Agrostis Stolonifera* (Copenhagan, 1816), pp. 11–16, 21–7, 29–30; J. C. Drewsen, *Om Krybehvene, Agrostis stolonifera (engelsk: Fiorin)* (Risbenhavn, 1818), pp. 4–5, 21, 23–4. I am grateful to the Royal Library of Denmark for copies and to Dr Petri Mirala of the University of Helsinki for translations.
156 Richardson to Greenough, 30 October, 26 November 1817, 22 January 1818, UCL, Additional Greenough papers, 1501, 1502, 1503.
157 W. Richardson, *Memoir on the cultivation of fiorin grass drawn up by desire … of his imperial highness the Archduke John of Austria* (London, 1818), pp. 111–23, 129, 142, 146, 161; Frederick de Conynck, *Afhandling om fiorin gräsets odling; författad på begäran af och till underwisning för hans kejserliga höghet, ärkehertig Johan af Österrike* (Stockholm, 1818); *Farmer's Magazine*, vol. 75 (August 1818), pp. 346–8.
158 J. Donaldson, *Agricultural biography* (London, 1854), pp. 107–8.
159 P. J. Jupp, *The governing of Britain, 1688–1848* (Abingdon and New York, 2006), p. 106.
160 Richardson, *Letter on irrigation to the Rt. Hon. Isaac Corry*, p. 40.
161 Kidd, 'North Britishness', pp. 362, 364, 375.

5

Richardson and Malthus

Prior to the publication of Ricardo's *Principles of political economy* in 1817, the Reverend Thomas Robert Malthus was 'England's foremost political economist', who held the chair on that subject at the East India Company College.[1] Malthus's famous *Essay on the principle of population* (1798) argued that demographic growth would inevitably outstrip food supply and was controlled only by positive checks like famine, or preventive checks, which reduced the birth rate. A second edition (1803) attempted to ameliorate the bleak tone of the 1798 essay by emphasising delayed marriage and moral restraint, accompanied by education, in controlling population. By the time Richardson wrote his critique in 1811, two more editions of the population essay had appeared, along with Malthus's economic writings about high food prices (1800) and articles in the *Edinburgh Review* (1808 and 1809) on the problems of Ireland.[2]

Richardson's fiorin project had promised to increase food supply, but he claimed that the frightening implications of Malthus's population essay necessitated 'obtruding my sentiments on a subject out of my line' and 'hazarding my speculations on political economy'.[3] The resultant essay on Malthus is a long production (fifty-six pages) addressing famines and food scarcities, the English poor law and Malthus's views on Ireland. This complex work is sometimes penetrative and revealing, at other times repetitive and digressive; yet it merits detailed analysis. Richardson was the only Irishman to refute Malthus's views on Ireland,[4] and the economic aspect of his work has entirely escaped historians, though its importance was not lost on contemporaries. In 1816 Richardson's college friend, the Irish Attorney General William Saurin, forwarded his complaints about the misapplication of public money to the Chief Secretary, Robert Peel, who replied that as 'the economy is so much the order of the day', Richardson's views 'deserve attention'.[5] Moreover, as he wrote as an agricultural protectionist, Richardson affiliated himself to an important strand of contemporary economic thought.

Protectionism has been sidelined historiographically by a focus on free trade and decreased state intervention, which, it is argued, inaugurated the nineteenth century's defining liberal orthodoxy. Conversely, protectionism has been represented as the uncomplicated reaction of those with a vested interest in land who opposed any reduction of state assistance. However, recent research on protectionism's role in shaping a resilient conservative economic ideology recognises it as a multifaceted and sophisticated doctrine capable of reasoned and forward-looking economic analysis.[6] The implications of the fact that this Irish protectionist can be fitted into the broader protectionist debate have not even been considered. This chapter therefore examines Richardson's response to Malthus in this light. It argues that this angle of enquiry reveals sophisticated arguments about Ireland's economic position in the United Kingdom at a pivotal historical moment between the inauguration of economic warfare in 1806–07 and the end of the Napoleonic wars in 1815. Subsequent work by Richardson saw him trying to adjust his ideas in response to the post-war economic crisis. In all of this, however, he presented Ireland as an answer to England's economic problems, but his version of protectionism was inflected by his political Protestantism.

The genesis of Richardson's foray into political economy dates from May 1809, but he saw political economy as a subsidiary venture to fiorin, telling Greenough that the bogs business had 'thrown Malthus and Famines … quite out of my view'.[7] He cautiously sent Greenough drafts for comment before the Malthus essay eventually emerged, serialised in Vaughan Griffith's *Agricultural Magazine* for 1811 and printed as a pamphlet.[8] The essay was reprinted in 1816 with a new dedication, though still dated May 1811.[9] For ease of reference I have used this version. First, however, it is necessary to outline Richardson's arguments.

Reacting to Malthus

The essay began quietly, with a polite identification of issues for discussion and a recognition of areas of agreement. Like many political economists, Malthus saw the English poor laws as problematic.[10] Richardson agreed that the poor laws exerted 'intolerable' pressure on all ranks of society and also concurred on the related problem of famine occasioned by crop failures, which were unpredictable but likely to recur. Malthus was also right about the political danger of food scarcity.

Recalling the food riots of 1800–01, Richardson stated that hunger could trigger mob outrages or be grafted on to political grievances by agitators to produce a revolutionary situation, forcing the government to curtail liberty. Turning first to the problem of scarcities or famines, he assumed that, in a stable population, food supply should be sufficient to meet demand. The problem of a grain surplus was addressed in Ireland by exportation and by distillation. Storage of the surplus from good years was one way of dealing with dearth, but this response was problematic given the unpredictable cycle of harvests and farmers' resistance to expend capital for uncertain returns. The alternative, food importation, would give hostile powers the opportunity to cripple Britain. The only viable option, as Malthus and Arthur Young had stated, was to provide a substitute food during scarcity.[11] But what substitute?

Richardson believed that the potato was the ideal emergency alternative to corn. It had the necessary qualities of being wholesome, indigenous, always available and resistant to inclement seasons. In Ireland this vegetable was used as an agent of improvement, being capable of renovating exhausted land and bringing virgin wasteland into cultivation. The problem in England, as Richardson saw it, was how to create sufficient 'consumption' or demand to make potato cultivation viable all the time, so that it could be diverted to food substitution in times of scarcity. Drawing on his own twenty-three years of farming experience he outlined how, in normal years in Ireland, alternative 'consumption' above farmers' domestic requirements was found by feeding the surplus to cattle and horses. In bad years, like 1800–01, when prices rose, this surplus could be sold and the proceeds used to relieve the poor. He agreed with Malthus that potatoes should not be introduced wholesale to England, and recognised the dangers of overreliance on that crop in Ireland. However, he argued that, if the Irish example of constant consumption was followed, the potato was feasible as a substitute for bread in bad times. When potatoes were used as a bread substitute, livestock could be fed on hay or whins. The good health of the Irish showed that potatoes were nutritious, and J. C. Curwen's investigations showed that sufficient ground was available in England and Wales to produce the necessary quantities.[12]

Richardson understood that industrialisation had contributed to population growth in England and Wales and, though agricultural production was rising, it was not keeping pace. As was common knowledge, this deficit had resulted in Britain changing from a country which exported food to one which had become dependent on imports. The strategic dangers of this policy were well illustrated by the example of the West Indian colonies, where the population was growing. Until the

American Revolution, the West Indies had been largely supplied from the thirteen colonies. However, the independent United States now had the power of withholding provisions in exchange for concessions. A touch of country Whiggism appeared when Richardson bemoaned the fact that 'power was a dangerous trust' and that colonial officials in transitory office were often at the mercy of unscrupulous merchants. Assuming that potatoes were indeed a viable substitute in England, Richardson considered how their cultivation in sufficient quantities could be encouraged. The patriotic example of the public-spirited landed class would work down through the social scale, though middling-order farmers might mix patriotism with economic concerns. However, as Ireland again showed, potato cultivation could also be extended to 'coarse untouched ground' deemed to be waste. Here government intervention was required, in the form of premiums raised from the savings made in paying for imports, money which could be spent more safely on domestic production than in enriching hostile powers.[13]

Given Malthus's dictum that demographic growth would outrun provisions and lead to catastrophe, Richardson's advocacy of measures to increase the food supply meant that he had to address the highly controversial population issue. Instead, as Malthus advocated, of finding means to reduce population to the level of the food supply the country could normally maintain, Richardson contended that the supply should be increased to support the population. Indeed, it could be raised even further and enable the colonies to be provisioned from the mother country. Britain's population could thus grow even further, thus increasing its military manpower capability. Richardson quoted an unnamed American writer that 'War as waged by Bonaparte, is not now principally a question of finances, but of the resources of population'.[14]

Turning his attention to the English poor laws, Richardson agreed with those who saw them as counterproductive, including Malthus, for whom they were 'tyrannical' and 'unconstitutional'. Richardson produced a succinct analysis of their operation before advocating his 'remedy'. There were two sides to the problem: firstly, how to persuade people to accept a substitute in lieu of bread; and secondly, how to effect the necessary policy changes to encourage production of that substitute. In language reminiscent of Adam Smith's free trade argument, he noted how in times of scarcity, if matters were left 'to find their own level' without the 'interference' of the poor laws, the lower orders would either go short of their normal food or be forced to accept a substitute. However, the poor laws worked against this. By distributing corn and money, these laws obviated the need for the poor to economise and meant they would keep consuming at the same rate, thus driving up prices. The wealthy, believing they did good, wanted the poor to maintain their normal consumption. In an ideal world, it was acceptable that a 'moderate' proportion of wealth should indeed go to the parish poor rate. However, gentlemen must 'not be led by false lights' into committing 'private injustice' and 'public mischief' under laws which would effectively 'change the persons who must bear the greatest weight of the calamity' (the wealthy). Richardson explained how the burden escalated up the social hierarchy by categorising the recipients of poor relief. Problems began at the bottom of the social scale. Those whose wages were too low to buy bread received parish relief to purchase this scarce commodity and so drove up the price of corn. The next category, those who could just about afford bread in normal times, now found it out of their reach. They also sought poor relief, driving prices even higher and impacting on the numerous 'meritorious class' above them, until an enormous financial burden eventually landed at the door of property owners. Yet this onerous expenditure did not address the problem of recurrent scarcity and actually created more paupers. Like other writers on the poor laws, Richardson worried about their detrimental moral effects on this 'meritorious' class, who could support their families in normal times, but were reluctant to 'pass the line' into pauperism, and had 'more foresight' than those who spent their poor relief in 'the gin shop'.[15]

Addressing perceived links between dependency and dissipation, Richardson cited the English social reformer Patrick Colquhoun, whose social table for England and Wales was based on the 1801 census and a survey of paupers from 1802–03.[16] Richardson agreed that dependency led inexorably to 'vice' and cited parliamentary statistics since 1776 to prove that, despite the accelerating cost of poor relief, 'the power of money' only made things worse. Such empirical evidence proved that 'our remedy has increased the disease … corrupted their morals and increased their numbers tenfold'. The cure was the use of potatoes as a substitute food, as outlined in the first part of his essay. As the immediate cause of the problems was the absolute want of food, governmental policy should concentrate on introducing potatoes as a substitute and so reverse the upward spiral of cost, as this would halt the problem at source, in the lower levels of society.[17] But how and when could this be done?

Malthus understood the danger that starving people might confuse the right to charity with political rights. If potato substitution was left to times of hunger, then implemented suddenly, it might trigger disorder. Richardson advocated that it be introduced gradually in normal times, though he took the first of a number of patriotic swipes at Malthus, saying that Ireland had largely avoided the dangerous food riots England had experienced in 1800–01. Regarding the fundamental issue of how to encourage farmers to grow potatoes in the first place, Richardson had a novel solution. Earlier in the essay he had used the phrase 'unskilled interference' in relation to the existing poor laws; it now became clear what he meant by this. He still advocated government interference to replace the poor rate with a 'poor fund' raised by taxation; but, understanding that taxation was a protectionist bogey and that the country was taxed to capacity to maintain the navy and army at wartime levels, he cleverly conceived a permissive element. Arable land was to be taxed at one shilling per acre, but farmers could exempt themselves by growing an acre of potatoes for every twenty-five acres under cultivation. Having presented Ireland as the answer to English problems, Richardson now turned to the 'unpleasant subject … of Mr Malthus's observations upon Ireland'.[18]

Malthus had referred very negatively to Ireland without having visited the country (until 1817).[19] Richardson's objections to Malthus's depiction of 'my country … in the most odious and disgusting points of view' were partly methodological – he had pronounced on a country he had not seen – but they were also patriotic and political. Malthus argued that dependence on the potato led to early marriage and population growth, and the only brake on population outstripping food supply was the negative check of sexual restraint, something the Irish, who 'propagate like brutes', were incapable of. If England and other European countries exhibited impoverishment, why should 'rags and squalid poverty' be a uniform and defining characteristic of Ireland which distinguished it from other nations?[20] Richardson accused Malthus of utopianism in thinking that human nature could be adjusted to remove poverty and wretchedness. He also challenged Malthus's representation of the Irish as an indolent people whose country, despite its low-wage economy, could not produce manufactures for export as cheaply as England. Possessed of census data for England and Wales, Malthus believed that little was known of the size of Ireland's population, but Richardson used hearth-money returns to compute it at five million. He fundamentally disagreed with Malthus over the effects of population growth. Was an increasing population a resource or a danger? And did Ireland's population outstrip its resources? Richardson took a positive view and, claiming that Ireland's population was already half that of England and Wales and triple that of Scotland, speculated that Malthus actually feared the strength of this population in comparison to the rest of the United Kingdom. Gleefully resorting to statistics, Richardson argued that Ireland's import figures proved that it was an opulent country and that, unlike England, was a net exporter of grain. The essay's final paragraphs asked rhetorically why Malthus had poured 'such a torrent of abuse upon Ireland' and, more menacingly, asked where he had got his information from.[21]

Protectionism and political economy in post-Union Ireland

To analyse its significance, Richardson's 'open declaration of war against *Famines* and *Poor Laws*'[22] must be seen in the wider context of Ireland's post-Union economic position in the United Kingdom, economic warfare with France and the longstanding problem of Britain's dependency on imported food for a growing population. An Act of 1806 had removed restrictions on Irish grain-trading with Britain. The 1813 Select Committee on the corn trade noted 'the ability of Ireland to become eminently serviceable to this country by affording to its great manufacturing population abundance of food at moderate prices'.[23] This integration of Irish agriculture into the United Kingdom's domestic food supply was considered a consolidating benefit of the Union. It was embraced by many English protectionists, though some rejected what they saw as an 'Irish fix'. Nevertheless, the 1815 Corn Law was framed around the autarchic premise that national self-sufficiency in food was possible.[24]

After the war, many economists saw Ireland in Malthusian terms, as a country where population increase outstripped capital formation due to a backward agricultural system characterised by subdivision.[25] Previously, Richardson himself, along with witnesses to the 1813 Select Committee and the Farming Society of Ireland, agreed that Irish exports were growing as its agriculture and the social condition of the people steadily improved.[26] Therefore Richardson wrote against a background where the economic situation offered opportunities for Irishmen of his ilk to influence government policy. The potential was great; but so too were the risks. The assimilating logic of Union pulled in a different direction from the nationally oriented scope of protectionism. Although protectionism was the dominant outlook in parliament,[27] a residual view persisted in some quarters that Ireland was different. This ignorance of Ireland, as Richardson would have seen it, wilful or otherwise, had other ramifications. At its worst, this lack of knowledge could reverse the trajectory of his economic ideas and lead to English solutions being considered for Irish political issues like Catholic emancipation. He believed that if the information he provided was heeded, Ireland's economic potential outweighed its problems. Richardson's intervention in the economic debate also had wider intellectual significance. He deliberately grafted his ideas on to those of key British protectionists, enabling us to evaluate the contribution of Richardson's analysis to the developing general conservative economic discourse.

In this respect Richardson's private correspondence with George Greenough again gives insights into his published work. His links with British protectionists can be seen from his use of the statistics of his friend John Christian Curwen, an agricultural innovator, who was at this time an 'advocate of high protection'. Curwen thought Britain could be self-sufficient in corn – an economic objective which inclined him to give strong backing to Richardson's fiorin ideas.[28] Curwen thought increased capitalisation of agriculture was also necessary, while Richardson believed self-sufficiency could be accomplished at little cost by an improved deployment of existing natural resources of plants and population.[29] Nevertheless, despite their differences of emphasis, Richardson's ideas can be located firmly within the terms of the protectionist debate.

Richardson's Malthus essay was reprinted in the *Pamphleteer* in 1816 and dedicated to Richard Preston, the Devon landowner and MP for Ashburton who had supported the controversial 1815 Corn Law.[30] Preston had invested heavily and successfully in corn cultivation and was trying fiorin on his estate. He had also published on the poor laws and the Corn Law in the *Pamphleteer* and was an 'outspoken champion of improved protection to agriculture'. In 1816 he attacked the Liverpool ministry in parliament and wrote pamphlets on protectionist concerns, demanding punitive duties on imported corn and accusing the government of financial mismanagement.[31]

The food problem in 1816 involved a glut rather than a scarcity, but Richardson continued to blame the extant poor laws. They were 'a grievance greatly aggravated by scarcity, and made more intolerable by abundance'. Deploring 'the calamitous state of our country', Richardson noted how Preston's recent pamphlet on the distresses of the landed interest echoed his own plans for reducing the pressure of the poor laws without 'altering an iota' of the code relating to those laws. Richardson's reprint, however, twisted the logic of the original 1811 Malthus essay, claiming in the dedication to Preston that the direct consequence of his plans for potato culture and bringing wastes under cultivation would be to increase feed for livestock rather than add to the surplus of grain and further lower the price. Although this rather limp argument undermined Richardson's ideas by making them applicable to contradictory situations, the fact that he consistently sought English protectionists' support and entered the debate on their terms is significant, as it represents an Irish input into the landed interest's common concerns.[32] Richardson certainly made some impression on English agricultural circles, as Sir Benjamin Hobhouse used his work as the basis for a protectionist address to the Bath and West of England Agricultural Society on 'the 42 millions expended in ten years for imported corn'.[33]

Anna Gambles has recently argued that there were two central planks to the protectionist case, a strategic imperative driven by the insecurity of the peace after 1815, and ideological concerns informed by Malthus's population theory.[34] Although Richardson wrote during the war, he anticipated later arguments on both of these mainstream issues. Perhaps because of his bitter personal experiences of rebellion in 1798 and 1803, Richardson repeatedly emphasised the danger of corn importation putting power into the hands of 'rivals or enemies', and even argued that Irish agriculture had imperial benefits in supplying the colonies.[35] Richardson's experience of utilising Irish manpower in the yeomanry helped him see population more as a resource than the threat he believed Malthus did. Both he and Curwen meant to 'break Bonaparte as a wheat merchant'. Greenough heard how 'Mr Curwen and I are *separately* encountering Malthus's position – *bring down the population to the level of the food*; we engaged each in our *own way*, to bring up the *food* to the level of the population'.[36] Richardson's views on population were similar to those of protectionists who queried Malthus's ideas before the latter had come down unequivocally in favour of agricultural protection in 1814–15.

W. T. Comber and George Ellis challenged Malthus over the detrimental effects of population growth, which they saw as a means of increasing wealth. Ellis even claimed in 1813 that Malthus's preventative check of sexual restraint went against religious principles, as there was no contradiction between human reproduction and the produce of nature.[37] Some zealous evangelical clerics also opposed Malthus's sexual check as infringing biblical injunctions to multiply. Richardson's position was similar, though it came more from his patriotic refutation of Malthus's claim that the Irish 'propagate their species like brutes' than from any religious standpoint. He dismissed the great political economist's concerns as unrealistic for considering 'man rather what he ought to be, than what he actually is'.[38] To protectionists of this stripe, self-sufficiency was the golden key to free the nation from Malthusian determinism. In this respect Richardson shared the interpretations of two very influential men. Sir Humphry Davy and Sir John Sinclair both believed that self-sufficiency could be achieved by state intervention.[39] But Westminster's intervention in Ireland could be hazardous for Ascendancy Protestants, as the 1806–07 'Talents' administration's support of Catholic emancipation had recently shown.

This was a serious dilemma for Richardson. To oppose state intervention would have contravened a landed and protectionist shibboleth and flown in the face of his own advocacy of parliamentary premiums; yet unequivocally endorsing it would open dangerous doors in Ireland. The poor laws created a similar dilemma for English protectionists, though for different reasons. They needed the principle of intervention to maintain the Corn Law, but disliked the practice as it manifested itself in poor relief. As Richardson told Preston, he did not seek to remove the poor law but to negate its damaging effects on the populace and the landed interest. He squared this difficult circle by coining the term 'unskilled intervention'. Its implicit opposite, skilled intervention, meant informed government involvement based on observable facts about human beings and human nature, hence his recommendation of a permissive form of taxation to replace the poor rate. This stance paralleled an important strand of economic discourse, held by Tories like John Galt and Whigs like Curwen. This argument rejected self-interested protectionism by the landed elite in favour of a fiscal policy whereby they took responsibility for helping reduce the national debt and maintain the military establishment through some form of property tax.[40] Such privileging of facts over theory marked another radical difference with Malthus.

Salim Rashid has noted how the methodology of political economy drew on older sciences. The Geological Society regularly discussed economic problems like the Corn Law, currency and the poor law. Some members, including Greenough, favoured a Baconian inductive approach, rejecting universal theories in favour of empirical observation of the facts. Ricardo also attended the Geological Society and inclined towards a deductive methodology based on *a priori* reasoning. At its heart Baconianism was utilitarian and Rashid has argued that economic exigencies led to a 'revised' and 'considerably *a prioristic*' Baconianism.[41] Richardson started with *a priori* logic to establish the causes of famines and the characteristics an effective remedy would need, but then switched to inductive reasoning based on empirical evidence from Ireland to prove the solution would work. The general thrust was decidedly empirical and inductive and therefore very different from the approach of Malthus, who began a friendship with Ricardo in 1811 following his economic articles in the *Edinburgh Review*. Indeed, until Malthus's 1814 and 1815 Corn Law pamphlets signalled his conversion to protectionism, he was an advocate of Adam Smith's doctrine of free trade. Malthus's preface to the second edition of his population essay stated that he had deduced his principles from the writings of David Hume, Robert Wallace, Richard Price and Adam Smith.[42] Though Richardson himself, perhaps for polemic purposes, occasionally used the language of free trade and talked of the poor being forced to accept substitute food in a crisis if things were left 'to find their own level',[43] the factual realities were always against such theories. The problem must be tackled inductively, working from the facts towards a solution, not basing the answer on a theory.

This methodological disagreement also helps explain why an Irishman addressed the issue of a poor law which applied only to England and Wales. Ireland had no poor law until 1838 and Scotland, like Ireland, relied on a more voluntary system of poor relief centred on the church and charity. Though Scotland was central in his fiorin campaign, Richardson studiously avoided mentioning that country in the context of political economy, except to say that the national pride of the Scots, like the Irish, would react strongly if subjected to the kind of derogatory comment that Malthus had made on Ireland.[44] Suggestions that the Irish lacked the 'decent pride' which made Englishmen reluctant to take hand-outs were met with evidence that Ireland needed no poor law and could rely upon the kind of paternalist voluntarism that saw only one-twelfth of Richardson's parishioners accept assistance in 1800–01. This interpretation was similar to that of Irish landlords who argued that the reality of agricultural improvement obviated the need to extend the poor law to Ireland.[45] Richardson feared that, for Ireland, working from theory to practice threatened political as well as economic consequences. Malthus had done just that.

In 1809 Richardson told Greenough about his 'late perusal of Mr Malthus', making it probable that he also read the *Edinburgh Review* articles of 1808, which contained even more on Ireland than the population essay. These articles were actually Malthus's reviews of a book by Thomas Newenham on Ireland's population and a pamphlet on Irish tithe reform. The Newenham review in particular expressed outrage at recent parliamentary debates on the Catholic question. As Ireland's population was increasing rapidly and Malthus believed that sexual restraint was the only possible check other than famine, he had asked how there could be progress and enlightenment if the lower orders were not 'freed from their political degradation' by Catholic emancipation.[46] As Richardson instantly spotted, Malthus was applying an analysis from his own version of political economy to Ireland on a matter which had been previously considered on purely political grounds. Richardson berated him for pronouncing on Ireland without empirical evidence, never having seen the country. A visit would reveal that manufactures were not as bad as claimed, and that Ireland contained decent houses as well as wretched cabins. The real point was that Malthus's impressions were bogus and the creation of 'a party' (the Whigs) which sought only political power. To this end that party told 'the lower orders of people, upon whom they mean to operate, that they are oppressed and miserable' and put the blame the government and the Ascendancy. Even if they did not mean to stimulate outrage, they irresponsibly misrepresented grievances to be able to claim that their measures would cure them.[47] This uncomfortable fact prompted Richardson to tell Greenough that he was convinced that Malthus's 'Billingsgate abuse of Ireland' was 'picked up among the Talents', and that it was 'well known which of them he is looking up to'.[48] Given his hatred of the entire Ponsonby connection, Richardson probably meant George 'Snouch' Ponsonby, the Irish Whig who led the party in the Commons after the Talents' fall in 1807.[49] It is significant that, while he argued for protection, it was from an Irish Protestant perspective, and intended to move well meaning Englishmen, including Greenough, away from Catholic emancipation. Indeed, the protectionists Richardson particularly courted, Preston and Curwen, were Whigs who favoured Catholic relief. Emancipation, for Richardson, would be an example of 'unskilled intervention' beside which the poor law would pale in comparison.

In the post-war economic crisis Richardson continued to make his presence felt in England, where poverty and rural unemployment were topical subjects. Arthur Young complained in his diary for 1816 that the government had no policy about 'the ruined state of agriculture', while 'multitudes were almost starving for want of employment even with cheap bread'.[50] In May 1817 the Board of Agriculture announced a large premium (£100) for the best paper 'on the best means of employing the industrious and unoccupied poor'. Richardson could not resist applying.[51]

Realising that he needed to 'keep sufficiently clear of fiorin, which will not be listened to in England', Richardson arranged for Preston to send material under his parliamentary frank to Whitmore & Fenn of Charing Cross, who subsequently published it as *An essay on agriculture*. This essay was dedicated to Curwen, who approved of it. Concurrently Richardson pumped Greenough and his other London friends as to exactly what the Board of Agriculture required, a surreptitious approach which 'irritated Arthur Young'. Everything was done in a rush. Richardson sought Greenough's opinion but the essay had already gone to press. He belatedly found out that the Board wanted applicants for the premium to remain anonymous. As usual, Richardson had written in his own name; but, on hearing this, hurriedly asked the publisher to conceal his identity and, for the first time in his life, adopted a pseudonym, Pan Orisibatos. However, having 'marked myself pretty strongly throughout the work', it was too late for major changes.[52]

Several authors, including Richardson, were considered as equally deserving the premium, which was to be divided. However, on an objection being raised that Richardson's essay, contrary to the regulations, had already been printed, it was decided to give him an honorary medal. The dedication noted that Curwen had supplied Richardson with information too late to be included in the original. This confirmed that Richardson and Curwen agreed in the abstract that population growth in England and Ireland was prodigious and unstoppable, that more food was necessary and that much ground grazed by cattle could be used for grain. Curwen thought that more capital had to be invested in agriculture, while Richardson argued that tillage could be increased by planting fiorin on wastes if livestock which currently grazed on good ground were moved to these areas. The large-scale cultivation of wastes would provide employment for the 'industrious and unoccupied poor'. This would address another problem exercising Curwen. Not only would the food be sufficient for the population, but a surplus would be created, which would allow export.[53] It all came to nought. Arthur Young perused twenty-five claims for the premium on the 'state of the poor, the causes of their distress and the means of remedying it'. Many ideas were 'visionary' and most advocated public works on canals and roads and other sorts of improvement; but the government ignored them all.[54]

Patriotism and political economy

Looked at from the deceptively simple angle afforded by hindsight, Richardson's economic works also deserved to be ignored. Ireland's accelerating population growth and the Great Famine gave the lie to his promotion of Irish remedies for English problems. There was a hint of *déjà vu* in his tacit hope that his arguments would open a backdoor to England for fiorin, which, cultivated in conjunction with potatoes, also promised to increase the amount of arable land. He did not press this: 'important as fiorin is on these occasions, I do not avail myself of it in argument, simply mentioning it, that I may not appear to have given it up'. However, engaging with a public celebrity like Malthus would certainly keep it in the public eye.[55] Yet, despite the self-serving utopianism and inconsistencies, if seen in the context of the Napoleonic conflict and the post-war crisis, Richardson's economic writings reflect a widely held belief that Ireland's untapped resources held the answer to many of the United Kingdom's economic woes. His readiness to enter the debate was partly driven by a patriotism which envisaged Ireland as an equal partner in the Union. Richardson's argument about the country's improving condition is similar to that of the Cork corn merchant, Gerard Callaghan, who told the 1813 Select Committee that 'the advancement of Irish agriculture may be dated from about the year 1782', the year of the Irish parliament's legislative independence.[56]

Yet Richardson's positive and patriotic talking up of Ireland concealed concerns that Westminster policy-makers would reject the Protestant analysis of their new economic partner's political situation. His Malthus pamphlet reveals an implicit fear that Westminster simply did not understand Ireland, through lack of information. He was certainly aware of his own ignorance of the political world beyond Ireland, as he allowed Greenough to determine the most opportune political moment to publish.[57] Whatever else he was, agricultural innovator, economic patriot and defender of Protestant Ascendency, Richardson was no doctrinaire physiocrat, wedded to the primacy of land over manufacturing.[58] He came from a part of Ulster which was the centre of the domestic linen industry and had connections, institutional and personal, with the rising commercial town of Belfast. To assess the significance of these connections we must now examine Richardson's contribution to development of provincial science.

Notes

1 B. Semmell (ed.), *Occasional papers of T. R. Malthus* (New York, 1963), p. 7; P. James, *Population Malthus* (London, 1979), pp. 213–44.

2 J. M. Pullen, 'Malthus (Thomas) Robert (1766–1834)', *Oxford dictionary of national biography* (Oxford, 2004), online edition, May 2008, www.oxforddnb.com/view/article/17902, accessed 3 August 2011.

3 W. Richardson, 'Simple measures by which the recurrence of famines may be prevented and the pressure of the poor laws greatly abated by a slight and partial change in our common agricultural practice', *Pamphleteer*, vol. 7 (1816), p. 164.

4 James, *Population Malthus*, pp. 146–7.

5 Richardson to Saurin, 22 January 1816, Saurin to Peel, 15 March 1816, BL, Peel papers, add mss 40211, fols 157–60, 161–7.

6 A. Gambles, *Protection and politics: conservative economic discourse, 1815–1852* (Woodbridge, 1999), pp. 25–6.

7 Richardson to Greenough, 18 May, 20 June 1809, UCL, Additional Greenough papers, 1424, 1426.

8 Richardson to Greenough, 15 June, 29 August, 14 November 1810, 25 May 1811, UCL, Additional Greenough papers, 1444, 1445, 1447, 1450, 1457, 1459; *Agricultural Magazine*, vol. 8 (January–June 1811), pp. 132–3, 275–6.

9 Richardson, 'Simple measures', pp. 175–215.

10 D. Winch, 'Higher maximum: happiness versus wealth in Malthus and Ricardo',

in S. Collini, D. Winch and J. Burrow (eds), *That noble science of politics: a study in nineteenth-century political history* (Cambridge, 1983), p. 72.

11 Richardson, 'Simple measures', pp. 160–6.

12 Richardson, 'Simple measures', pp. 166–76.

13 Richardson, 'Simple measures', pp. 180–8.

14 Richardson, 'Simple measures', pp. 188–90.

15 Richardson, 'Simple measures', pp. 191–8.

16 Hilton, *Mad, bad and dangerous people?*, pp. 126–8; Richardson, 'Simple measures', p. 193.

17 Richardson, 'Simple measures', pp. 193–5.

18 Richardson, 'Simple measures', pp. 198–201, 205.

19 James, *Population Malthus*, p. 146.

20 A. Digby, 'Malthus and reform of the poor law', in J. Dupaquier and E. Grebenik (eds), *Malthus past and present* (London and New York, 1983), p. 100; Richardson, 'Simple measures', pp. 205–8.

21 Richardson, 'Simple measures', pp. 206–14.

22 Richardson to Greenough, 20 April 1810, UCL, Additional Greenough papers, 1442.

23 *Report of the Select Committee appointed to enquire into the corn trade of the United Kingdom*, [184] HC1813, vol. 3, pp. 479, 482, 497.

24 Gambles, *Protection and politics*, pp. 27–9 and footnote 22.

25 P. Gray, *The making of the Irish poor law, 1815–43* (Manchester, 2009), p. 47.

26 *Report of the Select Committee appointed to enquire into the corn trade*, pp. 497, 503.

27 Gambles, *Protection and politics*, p. 27.

28 Hilton, *Mad, bad and dangerous people?*, p. 305; J. V. Beckett, 'Curwen, John Christian (1756–1828)', *Oxford dictionary of national biography* (Oxford, 2004), online edition, October 2007, www.oxforddnb.com/view/article/37334, accessed 19 August 2011; Richardson to Greenough, 11 November 1811, UCL, Additional Greenough papers, 1471.

29 Richardson, *Essay on agriculture*, dedication (to J. C. Curwen), pp. iii–vii.

30 Richardson, 'Simple measures', p. 164.

31 R. Preston, *An address to the fund-holder, the manufacturer, the mechanic, and the poor on the subject of the Corn Laws* (London, 1815); R. Preston, 'The poor rates gradually reduced and pauperism converted into profitable industry', *Pamphleteer*, vol. 11 (1813), pp. 551–68; Thorne (ed.), *House of Commons*, vol. 4, pp. 883–5.

32 Richardson, 'Simple measures', p. 160.

33 Richardson to Greenough, 6 February 1813, UCL, Additional Greenough papers, 1479.

34 Gambles, *Protection and politics*, p. 27.

35 Richardson, 'Simple measures', pp. 165, 182, 190.

36 Richardson to Greenough, 11 November 1811, UCL, Additional Greenough papers, 1471.

37 Gambles, *Protection and politics*, p. 28.

38 Richardson, 'Simple measures', pp. 162, 206; Hilton, *Mad, bad and dangerous people?*, pp. 336–7; S. Rashid, 'Political economy and geology in the early nineteenth century: similarities and contrasts', *History of Political Economy*, vol. 13 (1981), 731.

39 Gambles, *Protection and politics*, p. 28.

40 Gambles, *Protection and politics*, p. 39.

41 S. Rashid, 'Dugald Stewart, "Baconian" methodology, and political economy', *Journal of the History of Ideas*, vol. 46, no. 2 (1985), p. 245; Rashid, 'Political economy', pp. 726–44.
42 Pullen, 'Malthus'.
43 Richardson, 'Simple measures', p. 192.
44 Richardson, 'Simple measures', pp. 206–7.
45 Gray, *The making of the Irish poor law*, p. 51.
46 Richardson to Greenough, 18 May 1809, UCL, Additional Greenough papers, 1424; James, *Population Malthus*, pp. 149–50; Semmel, *Occasional papers of T. R. Malthus*, p. 18.
47 James, *Population Malthus*, p. 146; Richardson, 'Simple measures', pp. 205–18.
48 Richardson to Greenough, 18 May 1809, 14 November 1810, UCL, Additional Greenough papers, 1424.
49 Hilton, *Mad, bad and dangerous people?*, p. 202; D. Mansergh, *Grattan's failure: parliamentary opposition and the people in Ireland, 1779–1800* (Dublin, 2005), p. 259.
50 M. Betham-Edwards (ed.), *The autobiography of Arthur Young* (London, 1898), p. 466.
51 Richardson, *Essay on agriculture*, p. i; Richardson to Greenough, 26 May 1817, UCL, Additional Greenough papers, 1497.
52 Richardson to Greenough, 26 May, 2 August 30 October, 26 November 1817, 22 January 1818, UCL, Additional Greenough papers, 1497, 1499, 1501, 1503.
53 Richardson, *Essay on agriculture*, pp. iv–ix, 79.
54 Betham-Edwards, *The autobiography of Arthur Young*, pp. 470–1.
55 Richardson to Greenough, 15 June 1810, UCL, Additional Greenough papers, 1444.
56 *Report of the Select committee on the corn trade*, p. 497.
57 Richardson to Greenough, 29 August 1810, UCL, Additional Greenough papers, 1447.
58 Gambles, *Protection and politics*, p. 39.

6

Richardson and provincial science

William Richardson's 1808 *Memoir on fiorin grass* for the Belfast Literary Society was the spark which ignited an explosion of criticism from the town's radical intellectuals. Like 'lit and phils' in other growing towns, the Belfast Literary Society had a broad intellectual remit and heard papers on scientific, historical, literary and religious topics. Richardson was a corresponding member and, as well as publishing with the Society, delivered a paper on agriculture as a science.[1] This chapter examines Richardson's role within the Belfast Literary Society and his interactions with Ulster's broader intellectual life to identify reasons for the controversy he generated and to assess its significance regarding provincial science. By the early 1800s Belfast was unquestionably Ulster's provincial capital and the Richardson controversy reflects wider divisions over the role and custodianship of knowledge. In Belfast as elsewhere, the ownership of knowledge and control of its associational and institutional manifestations contributed to the shaping of civic identities. The links between provincial and wider, often metropolitan, knowledge elites was a vital ingredient in this shaping process. Yet such an important component of civic culture was contested by rival political interests and, when Richardson wrote, Belfast was emerging from the trauma and divisions of the 1798 rebellion. In order to understand Richardson's role we must first briefly establish Belfast's associational background and set it in the context of similar provincial towns and cities.

Belfast background

By the late eighteenth century Belfast was an important trading port which had also developed a cotton industry. The town's commercial importance had led to the construction of the Lagan canal in the 1780s, which linked the town to rural Ulster. Belfast's population increased rapidly, from 19,000 in 1801 to over 53,000 in 1831.[2] Presbyterians constituted the bulk of this population, and many of these people were prominent in the town's commercial and business community. In Richardson's time there was a tiny Catholic community, and a small but powerful Anglican group which dominated the political representation and looked towards the lord of the soil, Lord Donegall.[3]

Peter Clark has recently argued that a symbiotic connection existed between urban growth and the formation of associations in Britain, but claimed that lower levels of urbanisation and the small size of the Protestant community retarded the process in Ireland.[4] However, recent research demonstrates that Belfast actually had a very vibrant associational life in the eighteenth century and compared favourably to similar towns in the American colonies or Scotland. For example, a Reading Society was established in 1788, the precursor of the later Society for Promoting Knowledge.[5] When Wolfe Tone visited Belfast in 1791 to help form the first United Irish society he noted the town's vibrant sociability.[6] This vitality was undisturbed by the 1798 rebellion and Union. By 1811 Belfast had Galvanic, Medical, Historical and Cosmographical Societies, as well as its Literary Society, which began in 1801.[7] In this associational sense Belfast was like other urban centres in Britain and Ireland.

Provincial scientific and learned societies began in Britain in the 1760s, increased numerically in the late eighteenth and early nineteenth century and peaked around 1830. Towns like Manchester, Sheffield, Newcastle, Derby and Birmingham developed sophisticated associational structures catering for a broad range of intellectual, cultural and political activities. Examples include the Lunar Society in Birmingham (c.1768), the Manchester Literary and Philosophical Society (1781), the Derby Philosophical Society, the Leeds Natural History and Philosophical Society (1783) and the Newcastle Literary and Philosophical Society (1791). Similar Irish provincial societies also started life in the 1760s, increased in number during the 1780s and peaked in the 1820s, though Dublin exhibited a different pattern as associational formations there decreased after Union.[8]

These provincial societies often engaged 'itinerant' lecturers to talk on scientific topics, though Edinburgh's Philosophical Society hired lecturers to spread science to the Scottish provinces.[9] In Sheffield visiting lecturers and local 'savants' created a recognisable 'scientific community' which became institutionalised from the 1790s in an associational network bound together by participation in societies and correspondence. Sheffield's scientific community had monthly readings of papers, lectures and laboratory-based activities as well as a library and museum. Belfast's Academy taught science in a utilitarian context for improving linen manufacture and also had a natural history society and museum.[10] Displays of natural history specimens often originated in personal collections in private houses of connoisseurs, like the industrialist John Leigh Phillips of Manchester and Jonathan Salt, the Sheffield cutler.[11]

Though the formation dates, activities and general ethos of these societies seem to indicate a general cultural phenomenon, historians have noted how the precise form provincial science took was 'regionally inflected' by local economic, social and political conditions.[12] Manchester's Literary and Philosophical Society appealed to 'marginal men' isolated by commercial occupations and religious dissent, and its social function legitimised their aspirations to join the predominantly landed and Anglican cultural elite. In Sheffield, by contrast, the 'scientific community' of itinerant lecturers and 'savants' remained apart from the 'social elite', from whom they were distinguished by political affiliation (a tendency towards radicalism) and religion (dissenters outnumbered Anglicans). In other regions scientific associations did not correspond to any political or social divisions, while in older centres like Bath and Bristol new societies grew organically from established social elites.[13]

Historians also note the dynamic role these societies played in promoting 'strong, distinctive and increasingly bourgeois civic identities'. Specimen collecting tended to become institutionalised with the emergence of provincial learned societies and fed into civic pride. But these civic identities were fluid: as towns developed in this formative period, civic life could be 'made and unmade'.[14] Being 'regionally inflected', these societies certainly contributed to the nature of the particular civic identity they helped mould. But equally, given that membership was motivated by social at least as much as scientific aspirations, a provincial society's status could be materially enhanced by its connections to central and metropolitan science. Members of provincial societies often had parallel membership of metropolitan learned bodies. Richardson, for example, was a corresponding member of London's Geological Society.[15] Correspondence facilitated the development of provincial knowledge networks and could also connect them with metropolitan societies. Indeed, this form of communication particularly suited botanists, who could use it for gathering information and exchanging specimens. Publishing could link the centre to the provinces. The Royal Society's *Philosophical Transactions* offered central publication for provincial discussion.[16]

But tensions could exist between centrifugal and centripetal tendencies. Belfast's reputation was vitally important after 1798, and its civic identity and the control of knowledge became vigorously contested. The parties to this cultural conflict were a radical interest, drawn to matters of intellectual advancement after 1798, and a conservative coalition of moderate reformers and loyalists. Richardson became a key figure in this dispute, which crystallised around the Belfast Literary Society. This battle for the control of knowledge was fought on two fronts – education and print – and the Belfast Literary Society's opening publication was authored by William Richardson.

Culture wars

Richardson dedicated this 1808 memoir to his former Trinity pupil William Bruce, a founder member of the Belfast Literary Society. Bruce and Richardson had shared interests in the classics and geology. On a tour of the Giant's Causeway in July 1806 Bruce had dined with Richardson's friends the Gages of Rathlin Island, and inspected 'Dr Richardson's specimens of water in basalt' and his fossils discovered at Portrush.[17] Referring in his dedication to his 1807 Board of Agriculture memoir, Richardson stated that, had he known 'an agricultural subject came within the limits and plan laid down by the [Belfast] society, I should not have carried my speculations upon an Irish grass into another country'. Bruce was also the principal of Belfast's Academy and Richardson apologised for 'seizing the opportunity of claiming some of the credit my Eléve has so deservedly acquired while initiating the youth of Belfast in the principles of religion, learning and loyalty'.[18] To understand why this apparently innocuous and polite dedication proved so incendiary we must examine the Belfast Literary Society's origins and character.

Some historians have seen the Belfast Literary Society as a cultural proxy for radicalism. Belfast has been compared to Edinburgh for its intelligentsia having 'transferred the initiative from political into civic activism'. It has also been argued that the catalyst for this was shared Enlightenment impulses and the strong bonds between Ulster and Scotland, which manifested themselves in associations like the Literary Society and the Belfast Historical Society (established 1811). Presbyterians heading to Scotland for their education left Ireland with a liberal theology and 'a political activism bequeathed by the rebels of 1798', and then returned 'energised by Scottish ideas'. Many Literary Society members were educated in Glasgow and this experience, according to this analysis, 'sat comfortably alongside their tendency to political radicalism'.[19] However, though the Literary Society certainly contained some ex-United Irishmen, it is misconceived to see it as a cultural outlet for radical energies, as this entirely misses the complexity of Belfast's cultural life. A founder member, the botanist and political radical John Templeton, resigned in 1803 rather than associate with James McDonnell, who had signed a subscription for the arrest of the United Irishman Thomas Russell.[20] Bruce was minister of Belfast's First Presbyterian Church and Henry Joy, a newspaper proprietor and historian, came from a leading Presbyterian Belfast business family. These men were 'cultural and political heavyweights' who moulded the town's intellectual life. They were also Whiggish reformers who had shunned the United Irishmen and held a gradualist position on Catholic emancipation, unlike many radicals who wanted immediate change. In 1798 Bruce had adopted a non-Orange loyalism, joining the 'black cockades' of the Belfast Merchant's Yeomanry.[21] An 1804 vice-regal visit saw the town's yeomanry, which also included some ex-United Irishmen, proudly parade before Lord Hardwicke. In 1806 Belfast's Nelson Club was established and the Literary Society heard the poem 'Trafalgar' by its member the one-time radical William Hamilton Drummond, minister of the Second Presbyterian congregation.[22] Bruce, Joy and Drummond relished any opportunity, whether through military voluntarism or learned associationalism, to demonstrate publicly that Belfast's civic identity had changed. Therefore the Literary Society did not represent a cultural turn by the town's remaining radicals; it was actually a bulwark for Bruce and Joy to guard Belfast's new civic identity against radical encroachment. Consequently the Society's activities and publications were in the firing line between the rival groups.

Bruce had originally envisaged a society which fostered advancement of knowledge as an admirable end in itself and was open to wider influences. His 1828 presidential address to the Literary Society reflected on these origins. Before its establishment in 1801, Belfast had lacked 'a central point of intercourse among the literary men of the town'. The town's sociable soirées discouraged serious scientific or literary conversation. Bruce considered athenaeum-type societies in Bristol and Liverpool as a model. Some people wanted a debating club, but Bruce pronounced this type of organisation inappropriate for subjects 'of a different and superior character'. In reality, he wanted to avoid creating a forum for open political discussion such as occurred in the 1790s, when Francophile radicalism was the dominant tendency. Like similar British organisations, the Literary Society's membership was solidly 'middle class', including a medical doctor, James McDonnell, and Henry Joy. The Society was locally orientated and was a voluntary self-regulating body. Its twelve original members could select new members by ballot. Visitors could be introduced, if they lived within ten miles of Belfast. But it recognised the need for connection to the wider scientific and intellectual world. Joy amended the constitution in December 1801 to allow corresponding members if they were proposed and agreed by ballot. The intention was to build connections with men of science elsewhere. Bruce proposed Richardson as the first corresponding member, in April 1802.[23] The other corresponding members were a mixed bunch, selected for their cultural attributes rather than their political inclinations. They included ex-United Irishmen like Dr Whitley Stokes of Dublin and David Baillie Warden, a Presbyterian licentiate exiled in America, but also Dr James Curry of London and Dr Boisragon of Bath.[24] Like Birmingham's Lunar Society, the Literary Society met 'on the first Monday before each full moon' to hear papers on manufacturing, law, commerce, political economy, medicine, pure science, art, literature and classical learning. But, as Bruce wanted, it avoided 'theological controversy and the political questions of the day'.[25]

The Literary Society wanted to make its mark by producing a major civil and natural history publication on county Antrim, which was to include a statistical account, like the Board of Agriculture's surveys. Bruce reckoned that such a publication would make the Society 'the cornerstone of our future fame' and be a patriotic 'benefit to the country'. The information would assist philosophers and also be of use to farmers, chemists, divines, miners and geologists.[26] The patronage of landed gentlemen was sought. The President, Dr McDonnell, consulted Richardson, Bruce and the Reverend Snowdon Cupples, before he approached Earl Macartney to approve the content of questionnaires before they were distributed. The accompanying letter noted how 'every member of the society will feel himself obliged by your condescending to make any addition … and if we can persuade you to take any considerable interest in the subject we are not afraid of succeeding'. The sanction of the Sovereign of Belfast and Lord Donegall's kinsman, Stephen May, was also sought. This desire to have aristocratic approval for their activities was epitomised by McDonnell's recommendation of a member, James Craig of Carrickfergus, to Macartney as a fit person to assess Antrim's agriculture and transport infrastructure. Craig had all the requisite attributes. He possessed 'a handsome estate, has planted a good deal, and went to Scotland to see the Duke of Atholl's and Lord Haddington's woods lately'.[27]

Control of Belfast's intellectual reputation was a very public bone of contention. In 1806 the local newspapers were 'full of controversies', stimulated 'to ascertain the real state of Ulster' and in particular Belfast, to assess 'the existence of literary taste'. This issue had been the subject of much debate and 'had the good effect of rousing people from a fatal languor that portended the death of literature among the merchants of this opulent place'.[28] In 1807 the radical challenge was boosted immensely by the return from Dublin of William Drennan. When Drennan had left Belfast in 1791, the town's intellectual and political atmosphere was vibrant; but he was appalled at the changes he found on his return. He told his sister that while the Catholics were politically active, Belfast's Presbyterians, once the backbone of the United Irishmen, had become moribund and the life of the mind was 'rotting like their flax steeping in … ditches', while their 'vapid newspapers' displayed 'what is called the literary talents of Belfast'.[29] Bruce initially tried to rehabilitate his old friend by drawing Drennan over the threshold into loyal society and suggesting he join the Literary Society; but the radical poet declined, damning their publishing record: 'I hear they don't do much good anyway.'[30] He had a point.

In April 1808 William Hamilton Drummond admitted to the Literary Society's glacial progress on the Antrim survey. Although they had 'collected a great many valuable materials and proposed various plans … nothing effectual was ever done'. With criticism growing, further delay was not an option and the Society decided immediately to set about publishing their papers 'detachedly', as 'fasciculi'.[31] A recent letter from Richardson had announced his intent to publish a paper on fiorin grass, the cost of which the Society would cover. Bruce established a 'committee of revision' consisting of himself, Joy and McDonnell to oversee the publication process.[32] Given the opposition, it was crucial that the opening publication was authored by someone of note, allowing the Literary Society to seize the cultural initiative and attract other intellectual heavyweights. In April 1808 the Bishop of Dromore heard that the first fasciculus contained 'a memoir from our friend Dr Richardson; a relation of an aerostatic voyage to ascertain the operation of magnetism and the composition of the atmosphere at great elevations', as well as a treatise on bleaching linen. Thomas Percy, Bishop of Dromore and a leading intellectual, who knew Richardson, was collecting reminiscences of men of the previous generation and Bruce asked if he would contribute these and lend his patronage to the Society.[33]

If Richardson's dedication to Bruce reflected the underlying issue of civic identity, the memoir's content was provocatively patriotic. Belfast's radicals had previously appropriated Irish culture and identity in events like the Harpers' Festival of 1791. Drennan himself had written a highly political poem entitled 'Erin' in which he coined the term 'Emerald isle'.[34] Richardson threw his own Irishness in their faces. Fiorin was 'our Irish grass', despite the objections of English botanists: 'though I am not ignorant of the botanical characters and name, *Agrostis Stolonifera*, of this grass, I shall persist in calling it by its old Irish appellation, Fiorin Grass'. 'Surely', he asked, 'we, Irish, have a good right to insist on the name being retained by which this grass has always been distinguished in the country where its value … is now more fully established'.[35] Richardson was never averse to provoking controversy for publicity purposes but, given the political context, this was coat-trailing at its worst. The response was rapid and stinging.

The radicals were busy with their own publishing venture. John Templeton, Drennan and a radical linen merchant, John Hancock, began publishing *Belfast Monthly Magazine* in September 1808. It solicited 'original criticisms on new publications … of peculiar interest to Ireland' and considered 'science and literature as the true friends of toleration bound by no exclusive rights, system of monopoly, sect or party'.[36] This periodical immediately attacked the Literary Society for publishing nothing since its inception and its second issue carried a caustic review of Richardson's memoir.[37] As Templeton handled natural history in the *Belfast Monthly Magazine* it is highly likely that he authored this review. Though he had resigned from the Literary Society, Templeton remained on dining terms with Bruce and Joy. However, extraordinarily for two men whose paths must have frequently crossed, Templeton's journals never mention Richardson, apart from a bitter controversy over Linnaeus, and he was a botanist who disparaged fiorin grass, which he thought spread mildew in wheat, one of Sir Joseph Banks's preoccupations.[38]

Templeton's review attacked Richardson personally as well as for what he represented. It began by mocking Bruce's ambitions to put Belfast on the same intellectual footing as British towns. The Literary Society's overweening ambitions compared unfavourably with literary and philosophical societies elsewhere. The aborted Antrim statistical survey was something even 'the most learned and numerous bodies have deemed beyond the limits of their powers'. Reacting to Richardson's dedication, the reviewer sneeringly reminded him of his own loyalist and yeomanry activities, twitting him for 'attempting to subject the workings of genius to an almost military discipline'. As for originality, in a thrust at Richardson's appropriation of patriotism, his memoir was merely a reprint of previous memoirs to the 'Grand Jury of the county of Armagh' and the Board of Agriculture. A further jibe at Richardson's insistence on fiorin's utility conceded that it had some value, but whether this lay in introducing a lucrative new grass or illustrating 'the futility of speculative opinions when directed to agriculture' was open to question. Templeton, a member of London's Linnaean Society and the type of academic botanist Richardson despised, showed his scientific colours in attacking Richardson's insistence on using the Gaelic 'fiorin'. This would discredit Belfast by choosing 'local nomenclature … unknown to the generality of readers in the north of Ireland' and 'totally neglecting a scientific name'.[39]

Other Literary Society members received hostile reviews in the *Belfast Monthly Magazine.* Drummond complained that a review, 'which I suspect was written by Dr Drennan', of his translation of the Roman philosopher Lucretius's first book was 'rash and unfair' in parts.[40] However, this complaint was nothing compared to the Literary Society's furious reaction to the Richardson review. The then President, Henry Joy, immediately grasped the point that the review was 'severe on fiorin and also on our society'. It was saturated with 'gall' and 'acid', and was glaringly illiberal in deprecating the Society's publishing efforts when it should have encouraged them. Compared to the scholarly review of Drummond's Lucretius, this piece reminded Joy of the work of 'an empty schoolboy returning from Glasgow' full to bursting with new ideas to enlighten the world. The review was 'ill-natured and meanly composed'. Even if written by 'hirelings' who 'knew nothing of the people of Ulster and of Ireland', it reflected badly on the periodical's proprietors and was a disincentive to 'inquisitive men seeking the good of their native land'. It was, however, 'curious' in its allusions to Bruce. Joy believed that Richardson's dedication praising 'religion, learning and loyalty' had been tactfully 'couched' to avoid attracting malevolent comment. Yet the opposite reaction had occurred, prompting Joy to ask 'In which of those three does the sting lie? Or will it be found in all?' Richardson, however, welcomed the controversy, as any notice boosted his wider ambitions for fiorin and gave a pretext to reply. Believing 'that the Society and the Fiorin will thrive better under a state of war', he anonymously authored *The reviewer reviewed*.[41] This in turn was reviewed by the *Belfast Monthly Magazine*, which asserted that Richardson's claim to overturn the adage 'to make hay while the sun shines' was nothing less than self-publicising gimmickry and denoted his wilful ignorance of Linnaeus.[42] As Richardson had calculated, the heat generated by this local row ignited wider interest. The original memoir was reviewed in the new London-published Tory periodical the *Quarterly Review*, which was read, according to Richard Lovell Edgeworth, by 'most of our principal farmers'.[43]

Joy was convinced that the Richardson review, which charged the Literary Society with aiming beyond its abilities, was politically motivated. But was a statistical account really 'beyond the powers of the people of Ulster and of Ireland?' Though its members were knowledgeable about botany, mineralogy and antiquities, the Society would not demean itself by inviting 'assistance from *all* quarters, and solicit communications from the farmer and the mechanic, as well as from the scholar'.[44] The stalled Antrim project proceeded in modified form, absorbing some of the original contributors, including Templeton. Richardson had already prepared an account of the Giant's Causeway, but now had to divide the geology section between himself and McDonnell. More delays followed. Richardson complained that McDonnell had 'long promised the eastern side of Antrim and has *positively excluded* me from it', but was 'indolent' and had 'done nothing himself'. McDonnell understood mineralogy but his tardiness irritated the prolific Richardson.[45]

The much-maligned publishing project eventually emerged in 1812 as the *Statistical survey of the county of Antrim*, edited by John Dubordieu, Rector of Annahilt in county Down. It contained extensive contributions from Richardson which reflected his geological knowledge of Antrim basalt generally and also on zeolite and ochre, which were found in that rock. He showed his expertise on the Giant's Causeway by writing an itinerary for scientific visitors.[46] Richardson had a more positive attitude towards Dubordieu's survey than Bruce, who remained bitter about the failure of his original project to platform Belfast as a centre of learning. This reflected their different motivations. Bruce's civic aims contrasted with Richardson's personal ambitions. Bruce felt the Literary Society had fruitlessly pursued 'this phantom for a longer time than might have sufficed for composing an encyclopaedia', only to find their original idea was resurrected 'by a clergyman of the county of Down'.[47] Dubordieu's publication represented the miserable end of their aspirations to emulate British literary and philosophical societies. However, Richardson's ambitions meant he saw the Antrim publication as a means to a wider end. As the work neared completion, Richardson met Dubordieu and agreed that his contributions could also be separately published, to their mutual benefit. For his part, Dubordieu hoped that dispersal of offprints at J. C. Curwen's Workington Agricultural Society would boost sales of the Antrim book. Richardson used every opportunity to publicise his own projects and did so with his contributions to the Survey by mentioning fiorin and including his itinerary of bogs for scientific visitors, in which he attacked Rennie's dictum that peat soil was sterile and to which he added a postscript detailing a successful fiorin planting.[48]

It is clear, therefore, that the Belfast Literary Society actively sought Richardson's involvement in its projects. Belfast's intellectual and cultural reputation was boosted by his demonstrable connectedness to provincial societies elsewhere, by links to aristocracy in Britain and

Ireland and by his recognition in metropolitan scientific circles through George Bellas Greenough, Humphry Davy and Sir Joseph Banks. This symbiosis can also be seen in Richardson's use of his influence with Greenough to recommend McDonnell to the Geological Society. When Greenough revisited Ireland in 1811, Richardson and McDonnell greeted him on disembarkation at Donaghadee, and insisted 'we must be together among the Belfast amateurs'.[49] In a mirror image, radical attacks on Richardson and the Literary Society were envenomed by the fact that he personally represented a conduit for the interplay between provincial and metropolitan science. He was a barrier to their ambitions and a protective shell for Bruce and Joy. If they were unable to dislodge him by attacking the institutional frameworks of knowledge and their publication record, then perhaps he was vulnerable over the content and purpose of knowledge itself.

The Belfast Literary Society also had educational aims. Education was a topic of considerable public interest due to the establishment by the Talents administration of a commission to enquire into the general educational provision for the people of Ireland.[50] In early 1807 the Literary Society had applied for government assistance to establish a 'philosophical' institution similar to a government-funded body at Cork. McDonnell approached the Talents' Irish Chancellor, Sir John Newport, and boasted of the 'progress of literature' in Belfast. He employed glowing terms to describe Bruce's Academy and a public subscription library which included a mineral collection and some 'philosophical apparatus'. But Belfast lacked a 'philosophical institution' to educate its youth in 'belles lettres, chemistry, experimental philosophy, and natural history'. Though several members could lecture on these topics, they had no proper teaching rooms. Thus they could not coordinate their activities for the 'great advantage' of the country 'by diffusing taste for knowledge, morals and good government among all classes of people'. Newport's secretary replied blandly saying the Society did not meet the same criteria as the Cork Institution, 'which does not embrace classical education', one of Bruce's favourite topics. In May 1807, at a meeting of Bruce's Academy, the members agreed to send an appeal to Newport's successor, John Foster, by the new President, Snowdon Cupples. This disingenuously claimed that Newport had encouraged the formation of a society and Foster promised to look into it. But by October no reply had been received and the incoming President, Drummond, again asked for a response. He conceded that 'by the mention of classical schools in their former letter, they did not mean to insinuate a wish for support in any of the lower departments of learning'. Drummond assured Foster that they wanted to establish 'a Philosophical, not … an Academical Institution' and looked to him 'as a friend to Ireland', but the envisaged funding never materialised.[51]

Belfast's radicals also saw education as a way to promote their reforming agenda. According to Drennan, he and Bruce had agreed in the 1780s on the need for 'a College being founded in the North'. On the death of its first principal, James Crombie, Bruce had assumed the headship of the Belfast Academy.[52] Arguably this met the need for a grammar school, but the collegiate aspiration remained unfulfilled. Ulster's Presbyterian ministers went to Scotland for their training. There had been an unsuccessful attempt to meet this need locally by establishing a university at Armagh in 1795, at the same time as the Catholic seminary at Maynooth was founded. But this came to nothing. The renewed initiative for an academic institution came from radicals like Drennan, Templeton and a fellow ex-United Irishman, William Tennent, who viewed education as a harbinger of political reform. They wanted a non-denominational, non-partisan institution and solicited support from merchants who saw the advantages of a broad education, landed gentlemen, bishops and government officials.[53] Templeton drafted a document to raise a subscription for an institution which would 'establish a system of education … honourable to our town and neighbourhood'. He claimed that Europe was growing more enlightened, and small states competed to become patrons of learning. In London, 500 printing presses 'are daily giving to the world productions which spread knowledge to the most distant regions and bring to that splendid metropolis immense riches'. Belfast needed something similar. 'Scientific and literary knowledge' was indispensable to the 'professional man' and was the gentleman's 'brightest ornament'. However, parts of Templeton's heavily political original draft were toned down for the printed version, which emphasised the purity of science, unsullied by the grubbiness of Belfast's politics. A sentence in the original version anticipating universal gratitude at the foundation '*when all recollection of opposition to this institution shall be swept into oblivion*' (original emphasis) became the more anodyne 'Let not, then, the plant of SCIENCE languish in its infancy for want of those genial supplies to bring it to perfection'.[54] The subscribers hoped that 'the town may be a centre from which lessons of science may emanate, not only illuminating the scholar and speculative philosopher, but enlightening the husbandman, the manufacturer and the artisan'.[55]

The entire issue was cloaked in controversy and intrigue. Before the subscription meeting, Bruce, using Joy as a mediator, advised Drennan not to attend, as there was 'a combination against him, and such attendance might *revive* impressions [i.e. of disloyalty] amongst a superior class of people'. Drennan retorted that he would be 'distressed indeed if any attendance on such a business should clash with the friendship … notwithstanding our great difference of sentiment on many points of public concern'.[56] For his part, Bruce rejected overtures from supporters of the Academical Institution to join with them, insisting that Belfast already had an institution of advanced learning and, in any case, their management system was defective.[57] There was a failed peace-making attempt with the Academical Institution's founders, 'if they would drop the idea of establishing schools for reading and writing and confine themselves to a Philosophical Institution'.[58] 'Inst', as the Academical Institution became known, has been described as 'a statement as much as a school'. Though Lord Donegall gave land for the school at an affordable rent, the government was very wary. The rotation of office holders (which was what Bruce meant by defective management) reminded one official of the French revolutionary constitution, and the government insisted that the plan to allow subscribers of five guineas a vote in electing officer bearers was amended, and the threshold was raised to twenty guineas.[59]

The Academical Institution was successfully established in 1810 and a Collegiate Department opened in 1815.[60] However, the tensions had not subsided. That July, the *News-Letter* published a letter from 'a lover of my country and its Constitution' who had just adjudicated at the school's student presentations on the subject of Caesar. Mostly these reflected well on the school and the town, but the writer was outraged by one boy's theme, that 'a conqueror is no more than a licensed highwayman' and called attention to Wellington, the victor of Waterloo.[61] This contretemps epitomised the difference between the Belfast Literary Society's meditated 'philosophical' institution and its Academical rival. The question turned on whether learning should flow along established lines of authority or create its own authority. The first view is reflected in Richardson's praise of Bruce's hierarchically structured instruction of Belfast boys based on the 'principles of religion, learning and loyalty',[62] whereas the latter's democratic ethos recalled the Belfast of the United Irishmen. The construction of knowledge mattered as much as its function and this drew Richardson and Templeton into conflict over botany.

John Templeton has been depicted as a mild man whose political radicalism was tempered by his 'love of nature and humanity'.[63] Though his humanitarianism is beyond doubt, Templeton's journals reveal a zealous political activist. He attended meetings and wrote addresses in support of Catholic emancipation and fiercely rejected the London Hampden Club's parliamentary reform programme as insufficiently radical. 'Of what use', he railed, would reform of an English parliament be to Ireland, where people recalled 'the flames of our burning houses … the scourging, the executions [in 1798] before that fatal act of union deprived us of the voice of our countrymen'. Templeton also made radical interventions at Belfast Historical Society debates, demanding that every member express their opinion, as he himself did in the most forthright terms. In a debate on Cicero and Demosthenes, he contrasted the former's ability as an orator with the latter's patriotism. 'Throughout his life he [Demosthenes] appears to have been equally steady to his friends, and to his country, the friends whom he first adopted, and the cause which he first espoused, he forfeited his life rather than forsake.' The audience would have readily grasped the parallels being drawn with Thomas Russell and Templeton's resignation from the Belfast Literary Society.

In 1815 Templeton pasted into his journal an exchange from the *Belfast News-Letter* between an anonymous correspondent and Richardson over the Linnaean system.[64] The Swedish naturalist Carl Linnaeus had developed an internationally recognised taxonomy, identifying plants by their reproductive organs and classifying them under a binomial system with Latin names for genus and species.[65] Prior to Linnaeus's *Philosophia botanica* (1751) there was no agreement about how to name plants. Some botanists had focused on description of species, others on families of plants and their discussions were always linguistically complicated by a plethora of names from various languages. Linnaeus's emphasis on sexual organs had raised a controversy in the 1770s and 1780s, with some naturalists considering his names disgusting and his system prurient. Charles Darwin's grandfather, Erasmus, had struggled to find suitable euphemisms in *The botanic garden* (1791).[66] However, these controversies were over by the early nineteenth century and the Linnaean system was widely accepted, but not by Richardson.

Richardson's earliest publications had railed against scientific botanists obsessed with naming plants but blind to their utility. Though Richardson knew the Linnaean term for fiorin (*Agrostis stolonifera*) he opposed the classification system, complaining that 'the botanist enumerates to us at least 150 varieties of grass which he has classed carefully and he details to us minutely … but of their properties he takes little notice'. 'Who cares', he complacently asked Henry Joy, 'for the barbarous Linnaean jargon?'[67] Richardson yearned to go public on the issue, but Greenough consistently advised him against publishing his extreme views on the 'Linnaean School'. By 1815, however, Richardson 'could not contain' himself and his virulent views appeared in the *Newry Magazine* in lurid detail. Linnaeans were obsessed with systems of nomenclature and reproductive generation, and their master had misled his pupils. Francis Bacon had identified 'systems' as impediments to scientific knowledge. What, Richardson rhetorically asked, would he have made of Linnaeus's system, which emanated from 'conning over his strange subject matter'. To indulge in consideration of plants' generative faculties 'removed all sense of decency' and provided 'pictures of more disgusting obscenity' than the 'brothels of old Rome'. Botanists were seduced by a 'love of fame', which stimulated them to find new plants and produce 'cumbrous jargon', but if they learnt their properties agriculturalists could study the economically relevant species with no need for useless information. Until Linnaeans proved the practical utility of their knowledge, Richardson vowed to retain his provocative phrase 'Linnaean folly'.[68]

The letter Templeton pasted into his journal was signed by a medical doctor under the initials 'J.L.D.' It is highly probable that this was William Drummond's younger brother James Lawson Drummond, who had graduated in medicine from Edinburgh and returned to Belfast. He was appointed physician at the Fever Hospital in 1814 and received the Academical Institution's anatomy chair in 1819. The younger Drummond gave papers to the Belfast Literary Society on anatomy, birds and bats and helped establish the Belfast Natural History Society in 1821. In his letter 'J.L.D.' noted waspishly how the character of fiorin grass was firmly established, but that people's ears were 'wearied on the subject'. Richardson's voluminous publications had consumed more paper than would have been required for an analysis of perpetual motion. But the burden of the criticism was intellectual, not personal. Richardson misunderstood Linnaeus and claimed incorrectly that he had consigned plants into orders if they had the same number of pistils and into classes if they had the same number of stamens. Richardson's anachronistic resurrection of threadbare arguments made 'J.L.D.' wonder why, as 'great men had failed to tarnish the glory of Linnaeus', Richardson 'comes forward in the *Newry Register* [*sic*] … without starting one able argument, or shewing that he is acquainted with one item of botanical knowledge, [then] sets himself up as an open and avowed opponent of the Linnaean system'. 'Botany', he sneered, 'has received its death wound at last'.[69]

Never one to ignore a challenge, Richardson responded immediately, blaming the editor for publishing an anonymous letter from a coward driven by ulterior motives and 'inflamed with personal hatred'. He denied any connection between his fiorin advocacy and criticism of the Linnaean system, quoting a recent letter from Banks urging him to 'proceed with spirit … against the prejudices of unbelievers'. Jealous botanists would have liked to claim discovery of the grass and turned against it when he had staked his claim. To the charge that he had wearied the scientific public with a pile of pamphlets, Richardson let loose a volley of classical quotations. Then, clearly suspecting that Templeton's surreptitious influence moved his critic's pen, made a sly reference to the Academical Institution subscription. He also alluded to Templeton's lack of publications, saying that his antagonists had 'never blotted paper but when they had some grand speculation of raising money'. 'J.L.D.' replied denying any personal attack, saying he did not know Richardson, which was probably true, and acknowledged his merit 'as a geologist'. Interestingly, he noted 'from the whole tenor of Dr Richardson's paper, it is evident that he considers me an old and established enemy'. The real issue was science, which 'should be held inviolably sacred, and that neither reputation, years, talent, nor aught else, should *ever* be deemed a sufficient pretext ... for any man who would wantonly invade her empire and desolate her fair fields with ruin and confusion'.[70] Richardson, for once, was quiet.

The Belfast Literary Society heard papers on ballooning, electricity and meteorology and this has revealed that it functioned as a lightning conductor to earth dangerous radical ideas without damaging the classical structure of the 'Athens of the North'. It is equally obvious that the Society's veneration for Richardson, described several times as our 'honorary' member, was grounded on what he represented, which paralleled how they envisaged Belfast's intellectual and scientific life. In an important sense, this association was like those of Bath and Bristol, which gravitated towards traditional elites and landed society. This echoed the social rapprochement seen in yeomanry, with the learned papers replacing officers' commissions as the currency of cooperation. Richardson's memoirs to the Ulster grand juries, his personal friendship with Earl Macartney and his Trinity connections can certainly be seen in this light. Moreover, as we have seen, his views on Presbyterians had modified since the 1790s and would have now been congenial to Bruce, Joy and Drummond. In the Belfast Literary Society and the Nelson Club, linkages with elite landed proprietors offered a *via media* which elevated patriotism over partisanship.[71] Richardson's involvement with Davy and Greenough, his memoir to the Board of Agriculture and Royal Irish Academy papers made him a celebrated man with the kind of metropolitan connections in Ireland and Britain to boost Belfast's reputation.

Conversely, the lionising of Richardson was highly provocative and challenging. For radicals what he symbolised explains their dislike of him and those he associated with. Drennan once dubbed Bruce and Joy as 'the Irish Pitt and Castlereagh'. Where the Literary Society's moderates saw patriotism and prestige, their opponents saw conservatism and hierarchy in Richardson and his patrons like Abercorn, who had carried Pitt's coffin.[72] Richardson's claim to be the founder of the yeomanry infuriated men who habitually saw that force as an instrument of official coercion. This dislike of Richardson's elite connections helps explain the *Belfast Monthly Magazine*'s accusation that his memoir was a just reprint of that to 'the Grand Jury of Armagh'. As that memoir actually addressed the juries of Antrim, Armagh and Tyrone, this misrepresentation may have been intended to exploit Armagh's bad reputation since the notorious Orange outrages. Richardson himself doubted that he personally was the target, replying complacently to Henry Joy's horrified letter by saying that Bruce was the real object because of his loyalism. Referring to his dedication to Bruce, Richardson noted: 'We might have bandied our compliments and loaded each other with as much religion and learning as we pleased, had we kept clear of *loyalty*.' Unfinished business from Belfast's recent past was the real problem: 'You had once a spirit in Belfast which though … not now in action is still, though dormant, very irritable.'[73] Either way, the prestige of scientific knowledge became ideal political ammunition in post-Union Belfast. Yet, though politics helped shape the respective factions and their views of the nature and purpose of scientific knowledge, neither group was narrowly parochial. The Belfast Literary Society wanted to graft itself to the wider context of improvement throughout the British Isles and Templeton's accusations that Richardson was a boorish anti-intellectual were conditioned by concerns about how Belfast was viewed from outside. Indeed, the venom with which Richardson was attacked was a back-handed compliment to his links to influential metropolitan knowledge elites.

It would be over-simplistic to characterise early nineteenth-century Belfast's intellectual and scientific activity as being shaped solely by politics. There were genuine ideological and cultural differences at work. Richardson understood that urban centres like Belfast, Dublin and Edinburgh were a market for the agricultural produce of their hinterlands, but his orientation was always rural. Belfast was growing fast and industrialising and there are signs of a more general tension between the traditional and the modern and commercial. Bruce's elitist concepts were getting left behind. In December 1811 he noted that the question of whether the town should build 'a public library and newsroom after the manner of the Liverpool Athenaeum' or follow merchants 'intent on erecting Commercial Buildings' was 'a difficult point to settle'.[74] Richardson therefore represented a particular type of improver and a view of the proper end of knowledge. This was hierarchically structured, with aristocratic patrons leading by example. His rhetoric of amateurism implied resistance to academic encroachment and stressed that the elite were the proper custodians of knowledge and its purpose was practical and patriotic improvement for the public good. One of his last papers to the Belfast Literary Society, in June 1815, was on 'agriculture as a science'. The Board of Agriculture advocated scientific agriculture and was an agent of traditional landed improvement. A contrast may be drawn between Richardson's hierarchical and elite orientation and the Academical Institution's aim to bring the 'lessons of science' to the 'husbandman, the manufacturer and the artisan' as well as 'the scholar and speculative philosopher'.[75] The Swiftian adage the *Belfast Monthly Magazine* quoted against Richardson – that 'the man who makes two blades of grass grow where one grew before has more real merit than he who conquers kingdoms' – therefore dripped with condescending sarcasm for this survival from another age.[76]

Notes

1 Minute book of the Belfast Literary Society, LHL, p. 70.

2 A. Blackstock, 'Loyalist associational culture and civic liberty in Belfast, 1793–1835', in J. Kelly and V. Comerford (eds), *Associational culture in Ireland and abroad* (Dublin, 2010), p. 49; J. Agnew, *Belfast merchant families in the seventeenth century* (Dublin, 1996); J. Bardon, *Belfast: an illustrated history* (Belfast, 1982), p. 66; J. Bardon, *A history of Ulster* (Belfast, 1992), p. 203.

3 W. A. Maguire, *Living like a lord: the second Marquis of Donegall, 1769–1844* (Belfast, 1984), pp. 84–5.

4 P. Clark, *British clubs and societies, 1580–1800: the origins of an associational world* (Oxford, 2000), pp. 138–9.

5 E. Magennis, 'Clubs and societies in eighteenth-century Belfast', in J. Kelly and M. J. Powell (eds), *Clubs and societies in eighteenth-century Ireland* (Dublin, 2010), pp. 466, 480; J. Killen, *A history of the Linenhall Library, 1788–1988* (Belfast, 1990), pp. 7–8.

6 M. Elliott, *Partners in revolution: the United Irishmen and France* (New Haven and New York, 1983), pp. 22–3; Elliott, *Wolfe Tone*, p. 142.

7 E. Magennis, 'Clubs and societies in eighteenth-century Belfast', pp. 466, 480; J. Bew, *The glory of being Britons: civic unionism and nineteenth-century Belfast* (Dublin, 2009), p. 9.

8 S. Shapin, 'The Pottery Philosophical Society, 1819–1835: an examination of the cultural uses of provincial science', *Social Studies*, vol. 2 (1972), p. 311; A. Thackray, 'Natural knowledge in cultural context: the Manchester model', *American History Review*, vol. 79, no. 3 (1974), pp. 678–9; I. Inkster, 'The development of a scientific community in Sheffield, 1790–1850: a network of people and interests', *Transactions of the Hunter Archaeological Society*, vol. 10 (1973), pp. 99–131; R. Bayles, 'Understanding local science: the Belfast Natural History Society in the mid-nineteenth century', in D. Altis and C. Mollan (eds), *Science and Irish culture* (Dublin, 2004), vol. 1, pp. 144–6, 149; J. Adelman, *Communities of science in*

nineteenth-century Ireland (London, 2009), pp. 15–16; Killen, *Linenhall Library*, pp. 7–8.

9 A. Secord, 'Corresponding interests: artisans and gentlemen in nineteenth-century natural history', *BJHS*, vol. 27 (1994), pp. 369, 383–4; S. Shapin, 'Edinburgh and the diffusion of science in the 1830s', in I. Inkster and J. Morell (eds), *Metropolis and provincial science in British culture* (London, 1983), pp. 149, 169.

10 Inkster, 'Scientific community in Sheffield', pp. 99–100, 105; Holmes, 'Presbyterians and science', p. 545.

11 S. J. M. M. Albertini, 'Placing nature: natural history collections and their owners in nineteenth-century provincial England', *BJHS*, vol. 35 (2002), pp. 294–8.

12 S. Naylor, 'Introduction: historical geographies of science – places, contexts and cartographies', *BJHS*, vol. 38 (2005), pp. 6–8; I. Inkster, 'Introduction: aspects of the history of science and science culture in Britain, 1780–1850 and beyond', in Inkster and Morell (eds), *Metropolis and provincial science*, pp. 13, 33–4.

13 Thackray, 'Natural knowledge in cultural context', pp. 678–9; Inkster, 'Scientific community in Sheffield', pp. 99–131; M. Neve, 'Science in a commercial city: Bristol, 1820–60', in Inkster and Morell (eds), *Metropolis and provincial science*, p. 180.

14 Albertini, 'Placing nature', pp. 291–2; D. Eastwood, *Government and community in the English provinces, 1700–1870* (Basingstoke, 1997), pp. 58, 67.

15 D. P. Miller, 'The usefulness of natural philosophy', p. 192; H. B. Woodward, *A history of the Geological Society of London* (London, 1907), p. 269.

16 A. Rusnock, 'Correspondence networks and the Royal Society', *BJHS*, vol. 32 (1999), pp. 155–6, 168.

17 Journal of a tour of the Giant's Causeway, July 1806, NLI, Bruce MSS, MS20888.

18 Richardson, *Memoir on fiorin grass to the Belfast Literary Society.*

19 E. Patten, *Samuel Ferguson and the culture of nineteenth-century Ireland* (Dublin, 2004), pp. 35–6.

20 I. McBride, *Scripture politics: Ulster Presbyterians and Irish radicalism in the late eighteenth century* (Oxford, 1998), p. 210; Killen, *Linenhall Library*, pp. 46–7.

21 Bew, *Glory of being Britons*, pp. 32–3, 72; A. Blackstock, *Double traitors: the Belfast volunteers and yeomen, 1778–1828* (Belfast, 2000), pp. 24–30.

22 Blackstock, 'Loyalist associational culture'; Bew, *Glory of being Britons*, pp. 63–4.

23 Address by William Bruce, NLI, Bruce MSS, 20898; Minute book of the Belfast Literary Society, LHL, fols 4–12; J. Bew (ed.) *Belfast politics: thoughts on the British constitution by William Bruce and Henry Joy* (reprint Dublin, 2005), pp. 1–4; Bew, *Glory of being Britons*, p. 9.

24 *Historical sketch of Belfast Literary Society, 1801–1901* (Belfast, 1902), p. 7.

25 Minute book of the Belfast Literary Society, 19 April 1802, 22 December 1805, fols 12, 34, LHL; *Historical sketch of Belfast Literary Society*, pp. 4, 32, 157, 182.

26 Address by William Bruce, NLI, Bruce MSS, 20898.

27 Dr McDonnell to Lord Macartney, n.d. [c.June] 1802, PRONI, Macartney Papers, D2225/7/75, 76, 77A; Minute book of the Belfast Literary Society, 6 December 1802, LHL, fol. 15.

28 John Story, Belfast, to Robert Anderson, 17 February 1806, NLS, Advocates' papers, ADV MS22.4.10, fols 46–7.

29 Drennan to Martha McTier, 17 April 1807, in J. Agnew and M. Luddy (eds), *The Drennan–McTier letters, 1802–1819* (Dublin, 1999), vol. 3, pp. 594–5.

30 J. Jamieson, *The history of the Royal Belfast Academical Institution, 1810–1960* (Belfast, 1959), pp. 6–7; Bew, *Glory of being Britons*, pp. 11–12.
31 Drummond to Anderson, 6 April 1808, NLS, Advocates' papers, ADV MS22.4.16, fols 75–7.
32 Minute book of the Belfast Literary Society, 4 August 1806, 1 February 1808, LHL, fols 42, 52.
33 Bruce to Percy, 6 April 1808, in Nichols (ed.), *Illustrations of the literary remains*, vol. 8, pp. 413–14.
34 N. Curtin, *The United Irishmen: popular politics in Ulster and Dublin, 1791–98* (Oxford, 1994), p. 35 note 91; A. T. Q. Stewart, *A deeper silence: the hidden origins of the United Irishmen* (London and Boston, 1993), p. 189.
35 Richardson, *Memoir to the Belfast Literary Society*, preface, pp. 20, 28, 30.
36 John Templeton's journal, 4 May 1808, Ulster Museum.
37 *Belfast Monthly Magazine*, no. 1 (September 1808).
38 John Templeton's journal, January–February 1807, 11 October 1811, 6 May 1815, Ulster Museum.
39 *Belfast Monthly Magazine*, no. 2 (October 1808), pp. 130–1; Killen, *Linenhall Library*, pp. 28, 51.
40 William Drummond to Robert Anderson, 7 June 1808, NLS, Advocates' papers, ADV MS22.4.16, fols 79–80.
41 Joy to Richardson, 5 October 1808, Richardson to Joy, 17 October 1808, LHL, Joy MSS, ix.
42 *Belfast Monthly Magazine*, no. 4 (December 1808), pp. 377–80.
43 W. Richardson (anonymously), *The reviewer reviewed, or observations on a review of Dr. Richardson's memoir on fiorin grass* (Belfast, 1808); *Belfast Monthly Magazine*, no. 2 (October 1808), pp. 130–3; *Belfast Monthly Magazine*, no. 4 (December 1808), pp. 377–80; *Quarterly Review*, vol. 2 (May 1809), pp. 348–55; *Memoirs of Richard Lovell Edgeworth begun by himself and concluded by his daughter Maria Edgeworth* (London, 1820), vol. 2, p. 377.
44 Joy to Richardson, 5 October 1808, LHL, Joy MSS, ix.
45 Richardson to Greenough, 11, 17 January 1808, UCL, Additional Greenough papers, 1411, 1413; Address by William Bruce, NLI, Bruce MSS, 20898.
46 Dubordieu, *Statistical survey*.
47 Address by William Bruce, NLI, Bruce MSS, 20898.
48 Richardson to Greenough, 18 October 1811, UCL, Additional Greenough papers, 1468; Dubordieu (ed.), *Statistical survey*, vol. 2, appendix 33, pp. 101–2, 106–7.
49 Richardson to Greenough, 11, 17 January 1808, 25 August 1811, UCL, Additional Greenough papers, 1411, 1413, 1467.
50 *Memoirs of Richard Lovell Edgeworth*, vol. 2, p. 305.
51 Cupples to Foster, 5 May 1807, Foster to Cupples 9 May 1807, Drummond to Foster, 27 October 1807, PRONI, Foster papers, T2519/4/349, 350, 433; Minute book of the Belfast Literary Society, 2 February, 2 May 1807, LHL, fols 46, 48.
52 Drennan to Joy, n.d. [c.1809], LHL, Joy MSS, ix; I. R. McBride, 'Bruce, William (1757–1841)', *Oxford dictionary of national biography* (Oxford, 2004), online edition, January 2008, www.oxforddnb.com/view/article/3762, accessed 21 August 2011.
53 Jamieson, *The history of the Royal Belfast Academical Institution*, pp. 2–4, 6–7; McBride, *Scripture politics*, p. 211; Bew, *Glory of being Britons*, pp. 11–12.
54 John Templeton's journal, May–June 1807, Ulster Museum.

55 G. Benn, *A history of the town of Belfast* (London, 1880), vol. 2, pp. 102–3.
56 Drennan to Martha McTier, n.d. [September 1807], in Agnew and Luddy (eds), *Drennan–McTier letters*, vol. 3, pp. 627–9; Drennan to Joy, n.d. [September 1807], Joy MSS, LHL, ix.
57 Jamieson, *The history of the Royal Belfast Academical Institution*, pp. 6–7.
58 Drummond to Robert Anderson, 7 June 1808, NLS, Advocates' papers, ADV MS22.4.16, fols 79–80.
59 Patten, *Samuel Ferguson*, p. 7; McBride, *Scripture politics*, p. 212.
60 Bew, *Glory of being Britons*, p. 12.
61 *Belfast News-Letter*, 4 July 1815.
62 Richardson, *Memoir on fiorin grass*, dedication.
63 J. Killen, 'The reading habits of a Georgian gentlemen, John Templeton, and the book collections of the Belfast Society for Promoting Knowledge', in B. Cunningham and M. Kennedy (eds), *The experience of reading: Irish historical perspectives* (Dublin, 1999), p. 106.
64 John Templeton's journals, 19 January, 15 June 1813, 17 May 1814, 6 May, 21 August 1815, Ulster Museum.
65 J. Uglow, *The lunar men* (London, 2002), pp. 266–7; Holmes, *Age of wonder*, p. 8.
66 *Times Literary Supplement*, 24 March 2006, review by Jim Endersby; Uglow, *Lunar men*, pp. 385–6, 424.
67 Richardson, *Essay on the improvement of the great flow bogs*, pp. 4–6; Richardson to Henry Joy, 23 August 1808, LHL, Joy MSS, ix; Dubordieu (ed.), *Statistical survey*, vol. 2, p. 101.
68 Richardson to Greenough, 13 August 1815, ULC, Additional Greenough papers, 1491; W. Richardson, *On the insufficiency of the Linnaean school to improve our knowledge in the natural history of the vegetable kingdom* (Newry, 1815), pp. 1–8.
69 *Historical sketch of Belfast Literary Society*, p. 163; *Belfast News-Letter*, 15 August 1815; Richardson, *Linnaean school*, p. 4.
70 *Belfast News-Letter*, 15, 25, 29 August 1815; John Templeton's journals, August 1815, Ulster Museum.
71 W. Bruce, *The Christian soldier* (Belfast, 1803), pp. 6–12.
72 Hilton, *Mad, bad and dangerous people?*, p. 170; Malcomson, 'A lost natural leader', pp. 81–2; Bew, *Glory of being Britons*, p. 32.
73 Richardson to Joy, 17 October 1808, LHL, Joy MSS, ix.
74 Bruce to Robert Anderson, 6 December 1811, NLS, Advocates' papers, ADV MS22.4.16, fols 91–2.
75 Benn, *History of the town of Belfast*, vol. 2, pp. 102–3.
76 *Belfast Monthly Magazine*, no. 2 (October 1808), p. 133.

7

Conclusion: Richardson's historical significance

William Richardson was a complex and often contradictory man. Although he endured ridicule towards the end of his life, he was hard to ignore. Following his death, aged eighty, in June 1820 at Clonfeacle Glebe, the range and content of his obituaries bespeak this significance. The nationally circulated *Gentleman's Magazine*, to which Richardson had frequently contributed, praised this 'ingenious Clergyman' for his distinguished pursuits in natural history, and particularly for the zeal with which he brought fiorin to public attention.[1] *Blackwood's Edinburgh Magazine* also carried a glowing obituary. Richardson was 'well known to the literary world' for refuting James Hutton's geological theories, discovering fossils in basalt and his research into whyn dykes. Moreover, 'the agriculturalist will long remember the zeal with which he brought into notice the valuable properties of fiorin grass'.[2] Locally, the *Belfast News-Letter*'s obituary noted how the 'republic of letters and society at large' sustained an incalculable loss by his demise. This piece was written by someone who knew Richardson well, possibly William Bruce, as it praised his classical knowledge and 'capacious intellect'. Geology and 'practical agriculture' were indebted to the man who had received gifts of honorary membership from 'the farming and literary societies of the Empire'. This writer highlighted Richardson's polemic style: a 'clear, nervous and argumentative ... style of reasoning altogether peculiar to himself. He was seldom, if ever, overthrown in any of those controversial discussions into which he was frequently led by the boldness and singularity of his speculations.'[3]

Though Richardson was undoubtedly distinctive, his was not a unique voice. Other Irish scientists and innovators, like Richard Lovell Edgeworth and Richard Kirwan, contributed to the 'republic of letters' on various topics. Richardson's, however, was an insistent voice and, more particularly, an authentically northern one. The 'wars' that he fought with critics do not alter the fact that, for most of his publishing career, his ideas were taken very seriously by important figures. Indeed, omnipresence of controversy shows that he was not ignored. Richardson's writings are also significant because they combined scientific and economic ideas with an analysis of Irish politics and were addressed to a wide audience. This final chapter will consider the formative influences for his views, his publishing and dissemination methods, his stance on the proper role of government and his conception of authority – social, military and intellectual. It will also examine how typical Richardson was among provincial clergymen. Finally, his legacy is assessed by looking at how he was remembered and in the 'after-life' of fiorin grass. Cumulatively, this allows an evaluation of the contemporary and historical significance of this idiosyncratic man.

Two surprises await anyone researching this forgotten figure. The historical neglect of William Richardson is belied by the sheer amount and range of his publishing. Though his personal archive is lost beyond retrieval, I have attempted to assemble extant printed works like pamphlets, memoirs and periodical contributions. Inevitably, this process cannot be done with completeness. Nevertheless, not counting short contributions to periodicals, reprints or manuscript drafts of works, he published around fifty titles between 1801 and 1818. By any estimation this is a very considerable output concentrated in a relatively short space of time. Moreover, the places of publication range from Ulster locations like Belfast, Coleraine and Newry to Dublin and London. The topics covered are similarly wide-ranging, embracing politico-military affairs, geology, agricultural improvement, canals and political economy. The narrow chronological period in which Richardson wrote is also very striking. In his own phrase, Richardson did not 'blot a page' until he was over sixty. Regarding Irish politics, he entered the 'republic of letters' immediately after the demise of Francophile republicanism in 1798 and during the immediate working out of Ireland becoming a partner in the United Kingdom. The preceding chapters have marshalled this material thematically, but this concluding chapter looks at Richardson and his works in their totality. The timing and scope of his publications have biographical importance and also broader historical significance for what they reveal about the period.

We will never know precisely why Richardson first took up his pen to write his yeomanry pamphlet in 1801. Arguably his experience during the 1790s necessitated tangible face-to-face action without the reflective luxury of print. This experience included stormy meetings with parishioners in militia disputes, feasting local gentlemen to build up a social and political interest, active involvement with the Knoxes over the yeomanry and his enforced flight to Dublin. These circumstances could explain why all we have from Richardson prior to 1810 are his letters to Abercorn about the militarisation of the Orange societies and a few letters to the Attorney General on law and order. His yeomanry pamphlet of 1801 was his first publication and helps explain why he started writing.

William Drennan's sister, Martha McTier, offered a commentary on the context for the yeomanry pamphlet which is rather less than objective but nevertheless interesting. Referring to Richardson's 'running for shelter to Dublin' in 1798, she noted how he '*now* has courage to write facts he kept very quiet at the time and record himself a brave Churchman'. The waspish Martha put stinging words into the rector's mouth, as she imagined him as saying 'I will puff off the Knoxes and so get preferment'.[4] As Richardson was never advanced in the church, this opens the possibility that the denial of ecclesiastical preferment turned him towards writing. This interpretation would be consistent with what is known about Richardson as an ambitious man. However, it does not sit easily with other facts which indicate that all the necessary preliminaries were in place to prepare him to burst onto the publication scene in politics, geology and agricultural improvement. We know that he was an active experimental farmer long before 1806, when he put pen to paper to tell Lord Meath about his ideas. We know too that his lengthy summer sojourns at Portrush were spent actively investigating the Giant's Causeway and in communicating with the many important scientific visitors that it attracted. Previous contacts with Thomas Knox and Abercorn were essential preliminaries for a work on the yeomanry and there also was a contemporary political context. By 1801 Thomas Knox was estranged politically and personally from Abercorn but was trying (unsuccessfully) to negotiate to stand again for Tyrone.[5] Richardson, as we have seen, was an important intermediary between these men, and the possibility of a personally advantageous rapprochement may have moved him to write.

However, Richardson's political thinking cannot be divorced from his interest in agricultural improvement. Yeomen, like the corn or potatoes they farmed, were an economic as well as a military resource. It is notable that several of the men Richardson knew, Bishop Percy of Dromore and Lord Selkirk for example, were also interested in schemes whereby rural manpower could become a military resource.[6] Indeed, the yeomanry pamphlet's dedication notes that there were 'a number of claimants to the honor of being founders of the Irish Yeomanry'.[7] Given that he wrote just after the Union, arguably part of the broader imperative for his venture into print was a concern regarding the control of knowledge about Ireland. Richardson's need to put the record straight can therefore be contextualised against the wider backdrop of this critical and pivotal moment, when it was far from clear what Union's impact would be. *Castle Rackrent*, Maria Edgeworth's novel about Irish gentry life, was published around the time Richardson began writing. Her father, Richard Lovell Edgeworth, prevaricated about Union but eventually voted against it, and his daughter speculated what its effect would be on Ireland's identity.[8] Irish parliamentary representation was lifted suddenly from the familiar surroundings of College Green and dropped into the colder and wider political world of Westminster. Ignorance of Ireland was common among the metropolitan political elite.[9] Irish affairs had to compete for consideration with those of Britain, opening up the danger that ignorance, prejudice and misinformation could influence governmental policy on Ireland and the position of the landed aristocracy. His voluminous writings on fiorin grass can also be seen in this light.

Union, it is argued, spelt the end of the Irish aristocracy as a ruling class.[10] The immediate circumstances – the 1798 rebellion and the danger of French invasion[11] – did not redound to the credit of Irish parliamentarians and Ascendancy Protestants generally. Richardson's writings on agricultural improvement, in which he sought to present Ireland as an answer to Britain's problems, can also be seen as representing a desire among Ireland's landed elite to present Britain's new partner as an economic asset rather than a political or strategic liability. To say that Richardson offered the Irish landed class a way to rehabilitate themselves and integrate into the United Kingdom does not mean that he ignored personal credit for discovering and promoting the grass. But beyond this, as his anti-Malthus essay epitomises, Richardson was primarily an economic propagandist.

As his early correspondence with Greenough suggests, he made efforts to get the Protestant political case heard at Westminster, especially after the experience of the pro-Catholic 'Talents' ministry. His own politics, however, were more complex and he denied any party connections. The passage of Union coincides with a gap in Richardson's correspondence, though it may be reasonably assumed that he accepted it, like most of Abercorn's supporters and Isaac Corry.[12] But placing him politically is not straightforward. Though he offered to write in defence of the loyalist cause in the 1790s, the offer was refused.[13] In December 1807 he told Greenough that his pet hate was the 'Talents' ministry. Yet several key figures in his fiorin campaign, like Curwen and Preston, were Whigs who supported Catholic emancipation. The 'Talents' Irish Viceroy, the Duke of Bedford, was also a premier aristocratic improver whom

Richardson courted. Richardson's views on the Catholic issue changed. Referring to his position in the early 1790s and, probably reflecting Abercorn's current stance, Richardson said 'upon the great question of Emancipation (*then* properly applied) I was a warm advocate, and for a long time [afterwards] but their conduct satisfied me that the small minority who *then* opposed were right'.[14] On this question Richardson and Musgrave arrived at the same conclusion but from different premises. This is significant regarding the shift of the neo-conservative position to the mainstream of Protestant thinking. Musgrave supported Dublin Corporation's 'Protestant Ascendancy' resolutions of 1792 on historical grounds, arguing that Catholic treachery and disloyalty were 'timeless'.[15] Richardson, on the other hand, was present-centred, having witnessed and experienced the resumption of agitation for emancipation and the courting of the Catholics by the 'Talents' and their Irish Whig supporters, who he believed whipped up grievances to propel themselves into power. Richardson's pro-Ascendancy politics drove his criticisms of Malthus's views on Ireland. However, if he could convince English Whigs that Ireland under a 'Protestant' political disposition had economic benefits for the United Kingdom, so much the better. This thinking would have been consistent with his grandfather's true Whiggery. Indeed, Richardson's politics paralleled those of another loyalist and near neighbour, Colonel William Blacker, who detested 'modern Whiggery' because it had departed from revolutionary principles as they were understood in the early eighteenth century.[16] In one respect Musgrave achieved what Richardson desperately wanted, widespread recognition in Britain. Whereas fiorin grass may not have had quite the publicity-attracting qualities of rebellion and massacre, Richardson's efforts to disseminate his ideas are worthy of notice.

In Richardson's mind there was a natural connection between the methods of disseminating his ideas and the structure of social, military and religious authority. In an ideal world Richardson would have continued as he began, using memoirs dedicated to the nobleman whose patronage he sought, which included testimonies from other landed men, ecclesiastical dignitaries or military officers. This was common – the Highland Society required witness statements for prizes and premiums.[17] In this method of dissemination, the nobleman's intellectual prowess and social respectability legitimised and publicised the knowledge. The personal influence of an aristocrat or member of the political elite elevated the knowledge into an unassailable position. Richardson developed this approach in Ireland but, when he tried to replicate it for a British audience he had to modify his tactics radically. This indicates two things. Firstly, as with his evolving thoughts on geology, it shows commendable flexibility for a man aged nearly seventy. Secondly, and more importantly for his historical significance, his need to change methods reflects an imperfect cultural juxtaposition between the partners in the new United Kingdom.

Richardson instinctively preferred the traditional elite-based and hierarchical structure of authority to the 'hirelings' of agricultural magazines.[18] The structure of intellectual authority should conform to social patterns. Chaos reigned unless enlightened, educated and public-spirited men played their assigned role. God had created the natural world but its benefits were often hidden. It was the responsibility of intelligent men to find these secrets and reveal them to society. Even apparently unproductive bogs could become useful. But people had to listen to those whose birth and education put them in a position to know what was best for all. Richardson's parishioners who had the vestry book snapped in their faces, agricultural critics denigrated as bigots and dolts, and his son Marcus whose eagerness for marriage contravened parental authority, all committed the same offence of thinking for themselves.

Authority also derived from experience. This assumption is noticeable in his criticisms of the Irish yeomanry. Ulster's surplus manpower was negated as there were insufficient resident landed officers: corps were nominally officered by that bane of the eighteenth-century patriot, the absentee, while the sergeants were 'low people'.[19] Such social ordering also informed Richardson's views on associationalism. Freemasons were just about acceptable when organised under the Grand Lodge and Lord Donoughmore, the aristocratic grand master, but lower-class 'hedge masons' were anathema. Orangemen, he incorrectly believed, were brought under control by assimilation in a properly officered yeomanry force. The wreckers who had used the Orange name were 'the *dregs* and *refuse* of Protestants, mostly journeymen weavers'.[20] The lower orders, like bogs and wastelands, were improvable. Orangemen could be made more respectable by contact with the landed gentry. Catholics were not inherently rebellious and still less murderous, and they too could be brought to loyalty by proper management and the exercise of natural authority. He saw it as his 'duty' to advise his 'parishioners of the R. C. Community' who were, he believed, in great need of it.[21] This view of an organic, ordered and holistic society was underpinned by the scientific tradition which Richardson venerated. Quoting Bacon, he noted how, when man attempted anything it was partial, but 'Nature' made everything 'arrive at perfection at the same moment'.[22]

Social connections could also bestow authority. One of the main prompts to move Richardson from his traditional methods of dissemination into periodicals was George Bellas Greenough, whom Richardson met along with Humphry Davy in 1806. The Greenough friendship reveals much about Richardson and the broader context for his work. Greenough was a wealthy man, having been left a fortune by a grandfather who adopted him after his father's death. This legacy enabled him to live independently for the rest of his life. He attended Cambridge before going to the University of Göttingen in 1798 to study law. Lacking any professional or financial imperatives, Greenough became interested in geology while in Germany, visiting mines in the Harz Mountains. On returning to Britain, he made geological tours of Ireland, Scotland and Cornwall (where he began a lifetime friendship with Davy), as well as of the continent. He was a respected scientist embedded in the metropolitan associational world, being a fellow of the Royal Society and a member of no less than thirty-seven other learned bodies. In 1807 Greenough was a founder member and first President of the Geological Society of London. He has been described as a romantic geologist and, like Davy, his interests also embraced art and literature and both men included Samuel Taylor Coleridge among their acquaintance.[23] When Greenough first met Richardson he was aged twenty-eight and the rector of Clonfeacle sixty-six, which begs the question of why such a close friendship arose. Despite the age difference they were close intellectually. Geology was Greenough's predominant interest and, like Richardson, he was a 'stern empiricist', ill disposed to new theories. Moreover, Greenough's metropolitan connections gave Richardson access to the wider intellectual scene and compensated for his geographical isolation. But Greenough was more than a channel: he was also a filter, who could render raw Irish knowledge more acceptable to an English audience. Richardson utilised this by having Greenough read drafts of essays. This transaction worked two ways. Greenough, as a young but politically undecided MP attracted by utilitarian reform, wanted Richardson's first-hand knowledge of Ireland to supplement what he had picked up on his visits there. Indeed, the opening sentence of Richardson's first extant letter (of over 100) to Greenough began 'Dear Sir, Previous to my answering your several questions on the subject of Irish tithes…'.[24] This access to scientific networks was mutually beneficial. There was a *quid pro quo* for Greenough in Richardson's power of introduction in Ireland. His journal for 1806 noted that 'Dr. Richardson had written to the Bishop of Dromore that we [himself and Davy] would pay him a visit'.[25] Richardson also introduced Belfast geologists to Greenough, who, in turn, could link geologists from anywhere in the United Kingdom to those from France and Germany. This intellectual connectedness to Greenough, Davy and Banks was therefore of immense significance for Irish provincial science.

Arguably there was more than convenience at the heart of this relationship; there was real empathy and personal compatibility. Greenough, like Richardson, had lost his father early in life and may well have seen him as a kind of father figure.[26] A paternal tone is evident in the advice Richardson gave Greenough on handling Davy during one of his bouts of overwork-induced ill health: '*Poor Davy* ... the moment he is able to quit London, send him off with his gun, fishing rod, or anything but a book.' This was the voice of experience as he explained how he had suffered something similar himself when 'my early pursuits involved me in the same scrape'; but 'fortunately I had Davy's liking for fishing ... and what was more important I understood the disease, applied the remedy and rallied when others succumbed'.[27] Greenough remained a faithful correspondent and visited Richardson and his family several times. If Richardson was a father figure to Greenough, in many respects the younger man was an ideal surrogate son and perhaps the scientific heir he never had. Richardson depended on Greenough for political information and scientific advice; he also sought his help for his real and errant son, with tuition for examinations and asking him to use his influence at the Royal Military Academy to have young William withdrawn to avoid expulsion.[28] Greenough was urbane and diplomatic in handling difficult meetings of the Geological Society.[29] Richardson, on the other hand, had a fiery and impetuous personality.

This volcanic temperament had deep origins. He was an only child whose father had died young and whose worrying lack of physical growth led his mother to lavish attention on him. Indulgence in youth shaped an adult supremely confident in his own ability and intolerant of contradiction. His family's anxious dancing of attention on him can only have solidified these tendencies. Richardson's indulgence of young William can be seen as a replication of his own childhood. We know little about his College days, either as a student or as a teacher. However, his comment that he, like Humphry Davy, could fall victim to what is now called stress suggests a volatile temperament. Nonetheless, his antidote was fresh air and energetic outdoor activities, and his robust health and vigour show that the early concerns about his health were unfounded. The combination of anxious indulgence and physical and intellectual robustness helps explain the tone of his works. It was hard to distinguish between confidence and arrogance in his pamphlets, with their patina of learnedness and pedagogy.

Richardson and Greenough's chalk-and-cheese personality combination worked well. That that the younger man convinced Richardson to enter the market-driven world of London booksellers and periodical publication is beyond doubt. This process, on which he embarked reluctantly, brought out Richardson's very real natural talent as a publicist, but it was not an unmixed blessing. In part this suited his naturally combative personality (he once admitted 'I love controversy').[30] But his character predisposed him to personalise criticism. It is notable how in these 'wars' he projected himself onto fiorin grass, which became his 'protégée'.[31] Though friends warned about the dangers of his pugnacious attitude, he found it hard to curb it. But it was better to be criticised than to be ignored. When coat-trailing succeeded, it helped Richardson's fiorin knowledge overcome his geographical isolation. Belfast's Literary Society may have been horrified at the *Belfast Monthly Magazine*'s vicious review of Richardson's memoir, but he was delighted that the controversy reached the national audience of the *Quarterly Review*.[32] However, it was a risky strategy, which could backfire. Richardson's attitudes reflected a wider sense of heightened and nervous national pride among the Irish Protestant elite after Union. He could not comprehend why English farmers could not redeem their national pride by accepting his recommendations about fiorin. Those farmers instead reacted sometimes with a negative comparison to the Scots and sometimes by dismissing him as an Irish 'bog trotter'.[33] For them, fiorin was foreign grass; for Richardson, it was nature's parallel to legislative Union, if they only would see it. This ability to provoke a reaction came less from his relentless accumulation of facts or the classical larding of his prose than from his contemptuous dismissal of alternative views. This was best seen in periodicals, where his striking statements could trigger a chain of comment and riposte. It is a reflection too on his personality that the more contradiction he faced, the greater grounds he gave for it. These undignified squabbles were typified in his tussles in the *Munster Farmer's Magazine* and bitter rows with English farmers in the *Agricultural Magazine*. Beyond the personalities, friendships and rivalries, several important broader points emerge.

The Irish Union is usually considered with regard to its constitutional, legislative, financial or military ramifications.[34] However, one of its immediate effects was to create confusion, as the lines of authority and administrative structures were far from clear. Though Union eventually failed, in its first decades that was far from inevitable.[35] This combination of opportunity and obfuscation gives the immediate post-Union years an importance which is not fully reflected in the historiographical literature for the period, which concentrates on 1798. The uneasy administrative and political adjustments were paralleled by accommodations in the cultural and intellectual arenas. Though the 'republic of letters' comprehended Ireland and Britain before 1800, the new arrangements shifted the centre of gravity away from Dublin. At this distance in time, the implications are hard to see. Therefore the Richardson–Greenough relationship offers a rare demonstration of how an Irishman and an Englishman responded to the practical demands, uncertainties and ambiguities of the situation. Richardson's correspondence and some of his publications reflect a wider unease among the Protestant Ascendancy after Union that the political system in the new United Kingdom could foist on Ireland an executive which could never have controlled an Irish parliament. Westminster governance conjured another consistent concern: jobbery. Misuse of public money had long been a patriotic issue, but Union institutionalised British ignorance of Ireland and, as Richardson saw it, opened opportunities for exploitation which would tarnish the positive image he sought to present of the country.

The Irish Bogs Commission began in 1809 as a direct result of the priority being put on domestic food production, due to Napoleon's Continental System and Britain's retaliatory Orders in Council. Proposals to drain bogs were not new, but the 'war of resources' gave the matter urgency. Government agencies and influential individuals wanted a parliamentary commission to survey bogs and recommend the practicality of drainage. The Board of Agriculture was interested in reclaiming wasteland generally and supported an initiative on Irish bogs.[36] The Irish Chancellor, John Foster, a Board of Agriculture Vice President, drafted the proposal. He told parliament that half of Ireland's bogs would produce more corn than was ever imported into Britain, while the other half would yield enough hemp and flax for both countries, and parliament established the Irish Bogs Commission and granted it funds.[37]

Commissioners were appointed to initiate the surveys and nominate 'engineers' to execute them. As fiorin was an agent for bog improvement, Richardson sent Foster an 'elaborate plan without comment'; he was clearly yearning to be made a commissioner and so steer the project towards his own plans with the Board of Agriculture.[38] The government had difficulty in finding suitable people, and overtures were made to Richardson, who was 'sounded whether I would consent to be one of the commissioners'. He offered service 'consistent with *profession*, *habits* and *time of life*' and even claimed that the Lord Lieutenant asked him to be a commissioner, to which he 'cheerfully consented', though he never actually became one.[39] Yet, from the outset, Richardson had ideological reservations about squandering public money on a useless enterprise which he suspected was an electioneering scheme. These doubts emerged in a pamphlet written in 1809 but suppressed until the issue of his involvement was resolved.[40]

He determined to 'hold my tongue which is not easy', but vented his anger privately to Greenough.[41] As the Bog Commission began, he complained bitterly of jobbery. His investigations of the Bog of Allen with Lord Rosse had utterly refuted the Commission's 'wild nonsense', which was unfit for the noble aims of improvement. 'You give us *£5,000*', (the initial parliamentary grant), he told his MP friend, 'and our bogs would not be surveyed for £20,000'. Yet 'what information would the most accurate surveys give us?' Rather than salaried surveyors, reclamation should be 'the work of numberless individual proprietors'; otherwise, 'in *their* hands', it will be 'a mere bagatelle'. Parliamentary stimulus might help, but '*your* money will be swallowed in job', whereas inexpensive operations by landowners whose bog-lands were a 'gold mine' ripe for exploitation could continue after the public money was absorbed. He now accepted that he would 'not be named a Commissioner having talked rather freely about *apparent* jobs' but blamed the Union parliament. 'Under *your* law', Greenough heard, Commissioners would 'sit in committee in *Dublin* and appoint Surveyors, Clerks, *under salaries* … in other words to dispose of Mr Foster's patronage as he ordered'. The number of places would 'astonish' and 'your £5,000 will make a poor figure among these Harpies'.[42]

Richardson's fury increased in April 1810 when Greenough gave him advance sight of the first report. It was a 'consummate mass of ignorance and barefaced fraud', intended for 'patronage and job; clerks, artificers, ingineers [*sic*] metamorphosed into *secretaries*, under great salaries, all needy dependents or electioneering friends'. Turning a worthy objective, improvement, into a shameful job compromised possible future parliamentary grants for the genuine 'improvement of Ireland'. 'Be assured', he warned, 'every one of their side scouts the barefaced job, while the crowd pensioned under it exults openly in their salaries'.[43] Richardson's fears reflected wider concern that the Commission was 'a plan for enriching private individuals at the expense of the public'.[44] Fearing an 'Irish job' the Speaker of the House of Commons, Charles Abbot, became an overseeing commissioner. These worries were not groundless. Two of Foster's nephews were among the nine commissioners, while the engineers included his friend Richard Lovell Edgeworth and the Dublin Society's mining engineer, Richard Griffith.[45] In late 1811 Richardson's views became public.

The 'peat bogs' essay which Richardson had held back from publication was addressed to Lord Selkirk, who had visited him at Portrush.[46] He explained his delay in publishing it by claiming that 'under Sir John Sinclair's schemes I dare not offend at present'.[47] The essay was hot material and was refused by Harding of Pall Mall, but taken by the

Agricultural Magazine and also printed as a pamphlet, updated in the light of the second bogs report. Despite Humphry Davy's recommendation that the commissioners should 'appreciate the value and importance of my excellent friend Dr Richardson's ideas on the improvement of bogs, by cultivating on them the indigenous Irish grasses', this report was generally unfavourable.[48] Richardson's essay contained technical and utilitarian critiques plus ideological objections. Irish bogs were of immense depth, as the commissioner's report had stated, but this information had no utility for agriculture. The enquiry into the need for manures (a favoured topic in Davy's Board of Agriculture lectures[49]) to fertilise bog soil prompted Richardson to ask why surveyors were employed to discover what farmers already knew. Deep-drainage proposals were based on incorrect assumptions that Irish bogs were like morasses elsewhere. The commissioners' support for road building led Richardson to elaborate on the purpose of public funding. He was not opposed to grants for roads *per se*, but feared unscrupulous proprietors would appeal to the commissioners for self-interested ends. Premiums should cover landowners' expenses in conducting experiments for the public good. If successful, the landowner could continue for his own benefit and at his expense; but, when these improvements were made, new roads should be entirely publicly financed. In other words, public funds should follow voluntary self-financed improvements and not originate in proposals 'to expend *public* money on the speculation that *private* exertion will follow'.[50]

Richardson avoided personal attacks, but shifted the responsibility from the engineers to the 'framers of the bill', in other words, Foster.[51] The appendix to Richardson's essay criticised Foster obliquely by attacking the Scot, Rennie, who, with Foster's introduction, had boasted how, 'in consequence of publishing some essays on the natural History of Peat Moss, which met the approbation of the Board of Agriculture, I have been appointed to survey the bogs of Ireland'.[52] Richardson, 'impressed with an ardent zeal for the improvement of Ireland', knew Rennie was vulnerable to a patriotic critique. Even Edgeworth admitted that it would 'disgust every true Irishmen … to be told he is too ignorant to reclaim our *native* bogs'.[53] Richardson kept 'clear of *persons*', but decided that 'the business is a glaring job'. Yet bog reclamation itself was 'a fine field [if] we can hit the right mode of conquering it' by funding the landowner-initiated improvement he advocated.[54] Three parties were interested: the government, the great landowners and the inhabitants. The 'nation' could not attempt the task alone, as it would 'open a source for fraud, job, patronage and endless expenditure'. Nor could major proprietors be involved on their own accord, but could use their influence to encourage and provide means for the 'mass of the people' to convert wastes. Parliament should offer incentive premiums to be paid when the improvement was successfully completed, thus privileging local knowledge over professional expertise.[55] However, the arguments were unavailing. The third and fourth reports of the bogs commissioners contained Richard Griffith's devastating opinion 'that fiorin does not merit the high character it has obtained'.[56] Richardson relentlessly complained about commissioners incurring massive debts, but 'not a spade yet put into the ground'.[57] This attitude reflected eighteenth-century economic patriotism over any project involving public money, like inland navigation.

Canals had attracted Irish parliamentary financial support and were seen by landowners as key improvements. Construction of the Newry canal began in 1731, which linked Lough Neagh with the Irish Sea. The Tyrone Navigation, completed in 1787, connected Coalisland and its coal deposits to Lough Neagh via the Blackwater. The Lagan Navigation had connected Belfast to Lough Neagh in the 1780s, while Abercorn built a short canal to link Strabane with the Foyle in 1796.[58] The Lower Bann was the only outlet for Lough Neagh, the gateway to the Ulster interior. Richardson saw the Bann Navigation, like bog drainage, as another area properly within the remit of economic patriotism and requiring public-spirited aristocrats, not self-interested jobbers. In 1811 he addressed a memoir to the Marquis of Donegall, who, though heavily indebted, was exactly the right sort of person. He possessed extensive estates, was hereditary High Admiral of Lough Neagh, owned its fisheries and was patron of several parliamentary seats, including Belfast.[59] As an angler intimately familiar with the Bann, Richardson had the authority of local knowledge. His parish on the Blackwater also gave him a first-hand knowledge of winter floods which made Lough Neagh's tributaries inundate thousands of acres of farmland. Typically he tried to kill two birds with one stone, proposing a scheme simultaneously to stop winter flooding and to open the Bann for navigation.

While Richardson's preface solicited aristocratic influence in traditional terms, his rhetoric assumed knowledge of current issues like free trade and protection. He sought Donegall's influence at a time when parliamentary appropriations were being considered for Irish improvement, particularly inland navigation. Central Ulster was one of Ireland's wealthiest and most populous districts, with a thriving linen industry, but it relied on the Newry and Lagan canals for imported goods. Supply was uncertain as both canals could suffer from low water levels and Newry's monopoly of trade with the interior was unhealthy and needed competition. Coleraine, the town nearest the Bann's outlet, would benefit enormously if that river became navigable. Tacitly acknowledging Adam Smith's thinking, this development would not simply transfer the monopoly, but would 'give a stimulus to trade in general, and increase consumption to the benefit of all'. Exports and imports were seen as a measure of Britain's economic performance against other countries.[60] As in the anti-Malthus essay, Richardson stressed that Ireland had changed from a grain-importing country to an exporting one. The new navigation would boost exports by opening additional markets and would encourage farmers to invest in improvement. He emphasised the need to convince parliament of the scheme's financial feasibility, which compared favourably with other 'great public works', like bog drainage. On a familiar theme, he argued that utilising what nature had provided meant that expensive canal locks were necessary only from the Portna rapids downstream. Upstream, from Lough Neagh to Portna, a 'Chinese canal' of straight deep cuts avoiding sand bars would suffice and would relieve winter flooding. His geological knowledge showed that the Portna shallows were caused by the termination of basaltic strata and could be deepened by removing broken stones, again ameliorating flooding. This would not disturb eel fisheries on weirs otherwise vulnerable to any plan to radically deepen the shallows.[61] His ability to apply analyses from separate areas of knowledge, in this case geology, to a different topic would be described today as synergy or lateral thinking. For Richardson it came naturally, as he saw everything in Baconian terms, as an organic whole.

The memoir on the Lower Bann criticised a 'professional gentleman', Mr Townsend, not Richardson's Cork adversary, but an 'engineer' preparing a report for the Board of Inland Navigation. This body was bent on 'extending its domination to LOUGH ERNE' by proposing a canal to link it with Lough Neagh (eventually begun as the Ulster Canal in the 1820s).[62] Richardson's patriotic language equated a government-appointed body with political despotism and his preoccupation with Bann navigation led him to attack Townsend again in 1815.[63] His own amateurish presumption in addressing measures 'which a professional gentleman claims to have digested with much care' was excused by the fact that the 'question is a public one'. As an Irishman, Richardson was 'of a country in the habit of freely discussing the public measures in which it is interested'. Townsend's estimation of £100,000 for making the Bann navigable was a deterrent to public funding and his proposal for deep cuts was misguided. His plans were theoretical and based on assumptions that he dealt with a 'trifling stream', not the river which drained central Ulster. Richardson had 'spent so many days angling from a cot in this mighty stream that I cannot help looking on the proposal with astonishment.' 'Though not professing to be an engineer', the 'dictates of common sense' persuaded him that Townsend was wrong about Portna, had had only 'a cursory view' of the river, and his idea of a canal from Coleraine to Portrush was ludicrous. Like Musgrave's *Memoirs*, such nonsense betrayed Townsend's reliance on second-hand knowledge, as he had delegated the surveying to someone else.[64]

Drainage had long interested clerical improvers. Back in 1730 Bishop Francis Hutchinson had published a treatise on 'lowering the waters of Lough Neagh and the Lower Bann' and the Reverend Arthur Forde, rector of Lurgan, had written on the same subject in 1759.[65] That Richardson's writings on this topic can be seen in this context raises broader questions about his role and typicality as a provincial clergyman. Many Protestant clergymen were active improvers, sometimes to the detriment of their vocation, like the early eighteenth-century Wicklow cleric who 'loved his coppice as much as his bible'.[66] Other clerics saw religious proselytism and improvement as an integral part of the same civilising project.[67] A further point of comparison concerns clerical involvement in politics and, especially in the 1790s, law and order, activities ranging from pamphleteering to physical participation in the suppression of radicalism.

Richardson's role in the creation of loyalism and the promotion of loyal associationalism can certainly be compared with that of other clerics. Though not a magistrate or an Orangeman, Richardson bears comparison to the Marquis of Hertford's agent the Reverend Philip Johnston of Lisburn and his acolyte the Reverend Higginson of Ballinderry. These men were instrumental in organising loyal associations which were modelled on Richardson's Dungannon Association, which also evolved into yeomanry corps. Like Johnston and Lord Londonderry's agent the Reverend John Cleland, Richardson took a physically active part against radicalism by procuring informers and even enrolling himself as a special constable in 1795, whereby he assisted magistrates in executing arrest warrants and in deploying the military. Consequently, like Johnston and Cleland he was in acute danger of assassination.[68] However, while these similarities exist, Richardson was different, in that he gave the lead in organising loyalism, a lead which others followed.

He also shared the same political concerns as many Anglican clerics, but tended to express these privately to individuals rather than publish on them. His concerns about the growing power of the Catholic hierarchy were expressed in letters to Greenough, not, as another Tyrone cleric, John Anketell, did, in print.[69] Similarly, Richardson shared, or would have come to share, the concerns of the Reverend John Hales, rector of Killeshandra about the Catholic hierarchy. Hales, like Richardson, was a doctor of divinity and a Trinity College fellow but, unlike him, was a prolific writer on papal authority, Catholic loyalty and agrarian protest.[70] Again, Richardson's views on these topics inform only his private correspondence, and his defence of the Ascendancy position after Union was by surrogates in parliament rather than in the form of direct and named intervention in print.

Anglican clerics were prominent producers of topographical literature. John Dubordieu, rector of Annahilt, edited the statistical survey of Antrim to which Richardson supplied extensive geological information. Antiquarianism was an important aspect of this type of literature and many Anglican clerics were enthusiastic antiquarians. Of 241 subscribers of the Dublin Physico-Historical Society in 1746, no fewer than 101 were Church of Ireland clergymen.[71] Richardson's friend and fellow member of the Belfast's Literary Society Snowdon Cupples, vicar of Lisburn, was an enthusiastic antiquarian who also published a sermon defending Philip Johnston from attacks over his loyalism and Orangeism.[72] Though not an antiquarian himself, Richardson's interest in natural history was contemporaneously congruous with this discipline and there was considerable ideological and methodological overlap with antiquarianism. Edward Ledwich was another antiquarian cleric and a prolific writer on this and on classical topics.[73] He published several antiquarian studies of Ireland and edited one of the Dublin Society's statistical surveys of Aghaboe, in Queen's County.[74] Ledwich's associational links had a similar range to Richardson's, as the former was a member of the Royal Irish Academy, the Scottish Antiquarian Society and the Society of Antiquaries in London. Patriotism was another point of comparison. Antiquarianism and natural history mutually informed the topographical literature pioneered by Sir John Sinclair in Scotland and replicated in England and Ireland. Acquaintance with the nation's history and human and economic resources was seen as essential to maximising its potential and enhancing its reputation. Moreover, each field of enquiry prioritised fieldwork and required evidential proof for arguments.[75] Richardson's patriotic economic outlook and analysis were similar to those of Ledwich, who revered Sinclair's 'patriotism' and believed that economic information was vital as, even before Union, he feared Ireland was 'a country daily misrepresented by ignorant itinerants and whose internal state, wealth and resources are but very superficially known even by the more intelligent of our own countrymen'.[76] Such remarks anticipated Richardson's patriotic contempt for those who pronounced on Ireland after brief visits or, still worse, pontificated, like Malthus, without having set foot in the country. However, notwithstanding these similarities, despite the fact that he shared the views and dangers of clerical magistrates and even though his fiorin writings can be seen as part of the canon of clerical improving literature, Richardson was ultimately atypical in the lengths he went to and in the scope of his ambition. He was an agricultural improver with national currency. This privileging of improvement over his clerical duties was recognised in a backhanded way by the Anglican Primate, who threatened to remove him from his parish for subjugating religious concerns to promoting improvement in Scotland. This persisted. Richardson was still consistently engaging with political and economic issues until well into his seventies, thus raising the question of his legacy.

In the middle of the nineteenth century, John Donaldson's *Agricultural biography* (1854) delivered a damning verdict. It described William Richardson as a 'very learned person' possessed of a 'sanguine temperament'. He had recommended fiorin grass 'under all circumstances' but, though it worked in Ireland, it 'failed' in England because it produced poor amounts of hay, had 'inferior nutritional properties' and was difficult to sow. In debate Richardson combated all 'valid objections' with 'heated animosity' but 'possessed little practical knowledge and much less cool consideration', and his writings were 'hastily and carelessly concocted' and hence 'neglected'.[77] This judgement echoed into the twentieth century. Writing in the 1960s of English dairy farming, George Edwin Fussell said Richardson 'was an enthusiast but not a good scientist'.[78]

Donaldson was right about Richardson's personality; but, while many of his works had been dismissed by the 1850s, his network of acquaintances proves emphatically he was not neglected at the time. A non-exhaustive list includes two of Ireland's premier aristocrats, Abercorn and Downshire, peers like Aberdeen and Selkirk, and major British scientists such as Sir Joseph Banks, George Bellas Greenough and Humphry Davy. A measure of Richardson's legacy is the fact that fiorin grass did not fall out of use on the death of its promoter, particularly in Scotland. His struggles to convince the Highland Society of fiorin's credentials eventually bore fruit. Since 1812 the Society, wanting to test Richardson's claims, had offered a premium for whoever could grow the greatest quantity of fiorin on not less than two Scots acres. The premiums were advertised for several years and a decision made in 1816. Among the claimants were Richardson's key Scottish supporters General Sir James Stewart of Coltness, Norman Lockhart Esq. of Carnwrath, John Baird of Shotts Iron Works and Mrs Ann Trotter of Castlelaw. In 1820 the Highland Society published an account of several years of Scottish experiments. 'The original patron and champion of fiorin, Dr Richardson of Ireland' had permitted Stewart to pass his correspondence to the Society as the basis of a guide for the many 'extensive proprietors' interested in Highland improvement.[79] They offered a piece of plate worth twenty guineas or a money equivalent value for whoever could cultivate the greatest quantity of fiorin per Scots acre in Inverness, Argyll, Ross and Cromarty and the Islands.[80] Despite criticisms from the likes of Patrick Miller Junior, many who had tried fiorin kept using it. One of Richardson's arguments was that fiorin would boost population growth by opening up new areas for cultivation.[81] By the 1830s, John Baird's success with fiorin at Shotts Iron Works had helped the population of Cambusnethan parish reach 2,000. Baird had planted thirty acres of fiorin, which had 'for twenty years' been most productive and yielded 500 stone of hay per acre from ground previously 'worth nothing'. Alexander Young of Harbourn was 'advantageously' cultivating fiorin near Edinburgh and getting the same yield as Baird.[82]

Richardson's hunch that fiorin could alleviate rural poverty by making every tenant an improver also had purchase after his death. This derived from the enthusiastic reception Richardson's fiorin ideas had in the Low Countries. A benevolent society had been started in Holland to settle paupers in 'colonies for the indigent' on unproductive land, to be farmed to produce rye, turnips potatoes and barley. In each year of a four-year cycle they were to plant 300 roods of fiorin. Ireland's Society for Improving the Condition of the Lower Order of Tenantry and of the Labouring Population was interested in this model, as was the Highland Society of Scotland. In 1832 the *Dublin Penny Journal* quoted the Highland Society's investigation of the 'poor colonies of Holland' to recommend this example for Ireland's 'idle' lands like the Curragh of Kildare.[83]

But fiorin ultimately failed. One reason was that Richardson was wrong about its ubiquity. Curtis had raised the problem in 1813 and this vexed question was laid to rest in the 1820s. In 1824 James Townsend Mackey, a Royal Irish Academy member and associate of the Linnaean Society, read a paper based on his catalogue of Ireland's indigenous grasses, including distinct varieties of *Agrostis*. Mackey averred that *Agrostis stolonifera*, 'the famous fiorin of Dr Richardson', was distinct from *Agrostis alba*, unlike the Scottish botanist, Dr Hooker, who believed them to be the same species.[84] Experiments at the Duke of Bedford's Woburn estate had also identified several varieties of *Agrostis stolonifera*, hard to distinguish from true fiorin, and known, as Richardson's critics had always insisted, as couch or squitch grass. Climate and environment were also a deterrent. In 1825 John Claudius Louden's *Encyclopaedia of agriculture* carried a section on the cultivated grasses, including fiorin. Considering those to be used in hay meadows, it commented that 'agriculturalists … are not all agreed on the comparative merits of these grasses'. In scientific terms, despite the claims of Richardson's supporters, Humphry Davy's chemical analysis did not support fiorin. Davy had analysed the nutritive qualities of various grasses, including 'that singular grass fiorin', but found that it was not among the six best British grasses for dry and watered meadows. Davy reckoned that fiorin was useful only for 'particular soils and situations' and, even then, 'the opinions of cultivators' were not unanimous. Louden noted that 'few plants appear more under the influence of local circumstances than this grass'. On dry soils it was 'worth nothing', but 'if we may put confidence in the accounts given of its produce in Ireland', it was 'the most valuable of all herbage plants' on marl and moist peat or boggy soils. Davy himself had inspected the Countess of Hardwicke's experiments and analysed the nutritive content of fiorin hay, but concluded that, to be in 'perfection', it required a 'moist climate or a wet soil'. Moreover, there was no agreement on its qualities as feed for livestock. Lady Hardwicke's experiments revealed that cows, horses and pigs would eat fiorin hay, though another experiment, by the Duke of Bedford, comparing fiorin and conventional hay showed that horses showed no distinct preference for either type. Fiorin had limited merits in particular situations where it grew well, but it was never 'likely to be cultivated in Britain'.[85]

Richardson's old enemies, the seedsmen, gleefully supported this. In 1836 *The agriculturalist's manual* based its section on fiorin on a report by Peter Lawson & Son, seedsmen and nurserymen to the Highland and Agricultural Society of Scotland. This declared that 'the famous fiorin of Dr Richardson' seemed to be *Agrostis alba*, and that 'in this country [Great Britain] it has not in general been found to deserve the high characters which have been bestowed upon it' and it was 'better adapted for the climate of Ireland'. One reason for its partial failure in Britain was that those farmers who tried it did not do so on the peaty soil well supplied with running water which was best suited for it. Richardson's recommended mode of sowing cut-up stolons was 'tedious' and 'precarious'.[86] In 1843 a further publication by Lawson accepted that Richardson's Board of Agriculture memoir had 'created a considerable sensation among agriculturalists' in 1807 and, though the grass was known previously to English naturalists, Richardson at least had 'the credit of first acquiring for the fiorin a fair and general trial'.[87]

Some Irish farmers continued experimenting. During the Irish Potato Famine, the *Belfast News-Letter* printed a letter from a Donegal farmer to Ulster's Chemico Agricultural Society claiming a fine crop of fiorin from an acre of drained bog-land. But by 1853 the same newspaper's agricultural column advised farmers faced with 'sprits and fiorin which still infest the land' to 'exhaust every particle of the roots in order to extirpate them'. The disaster of the Famine reconfigured thinking about population and agriculture, making Richardson's comments on Malthus and his fiorin ideas seem hopelessly utopian and profoundly anachronistic. Yet his historical importance is not as an example of pre-Famine wrongheadedness, but as a unique provincial voice speaking to us from a formative period we know little enough about: post-Union Ireland.

Notes

1 *Gentleman's Magazine*, vol. 40 (1820), pp. 88–9.
2 *Blackwood's Edinburgh Magazine*, vol. 7 (April–September 1820), p. 584.
3 *Belfast News-Letter*, 23 June 1820.
4 Martha McTier to William Drennan, n.d. [1801], in Agnew and Luddy (eds), *The Drennan–McTier letters*, vol. 3, pp. 7–9.
5 Thorne (ed.), *History of parliament*, vol. 4, pp. 688–9.
6 Colonel Blacker's journal, vol. 2, pp. 41–2, vol. 7, pp. 75–6, Armagh County Museum; J. M. Bumsted, 'Douglas, Thomas, fifth Earl of Selkirk (1771–1820)', *Oxford Dictionary of national biography* (Oxford, 2004), online edition, www.oxforddnb.com/view/article/7920, accessed 24 August 2011.
7 Richardson, *Yeomanry*, p. ii.
8 G. Watson (ed.), Maria Edgeworth, *Castle Rackrent* (Oxford, 1964), p. 5.
9 T. Hoppen, 'An incorporating Union? British politicians and Ireland, 1800–1830', *EHR*, vol. 123, no. 501 (April 2008), pp. 328–50.
10 Malcomson, 'A lost natural leader', p. 85.
11 Jupp, *The governing of Britain*, p. 148.
12 Malcomson, *John Foster*, p. 168.
13 Richardson to Greenough, 20 December 1807, 3 January 1808, UCL, Additional Greenough papers, 1403, 1407.
14 Richardson to Greenough, 3 December 1807, 27 January 1808, UCL, Additional Greenough papers, 1402, 1417.
15 Kelly, *Sir Richard Musgrave*, pp. 14–15; Hill, *From patriots to unionists*, p. 221.
16 A. Blackstock, 'Orange songs in green books', in Blackstock and Magennis (eds), *Politics and political culture*, p. 75.
17 Anonymous, *Memorandum concerning accounts of experiments made in the cultivation of fiorin grass* (Edinburgh, 1820), pp. 1–2.
18 Richardson to Kater, 18 January 1813, Suffolk Record Office, HA 231/3/1/114.
19 Richardson to Kilwarden, 2 May 1803, BL, Harwicke papers, add mss 35739, fols 3–6.
20 Richardson to Greenough, n.d. [December 1807], 14 January 1808, UCL, Additional Greenough papers, 1401, 1412.
21 Richardson to Greenough, 3 January 1808, UCL, Additional Greenough papers, 1407.
22 Richardson, *Essay on the improvement of Dartmoor*, p. 142.
23 J. F. Wyatt, 'George Bellas Greenough: a romantic geologist', *Archives of Natural History*, vol. 22 (1995), pp. 61–71; J. Wyatt, 'Greenough, George Bellas

(1778–1855)', *Oxford dictionary of national biography* (Oxford, 2004), online edition, 2004, www.oxforddnb.com/view/article/11432, accessed 27 August 2011; J. Golden, *A list of the papers and correspondence of George Bellas Greenough* (London, 1981).

24 Richardson to Greenough, UCL, Additional Greenough papers, n.d. [December 1807], 1401.

25 Greenough's journal 1806, UCL, 7/13; Richardson to Greenough, 25 August 1811, UCL, Additional Greenough papers, 1467.

26 Wyatt, 'Greenough, George Bellas (1778–1855)'.

27 Richardson to Greenough, 20 December 1807, UCL, Additional Greenough papers, 1403.

28 Richardson to Greenough, 17 October 1811, 20 February 1813, UCL, Additional Greenough papers, 1469, 1482.

29 Wyatt, 'Greenough, George Bellas (1778–1855)'.

30 Dubordieu (ed.), *Statistical survey*, vol. 2, p. 38.

31 Richardson to Aberdeen, 20 November 1809, BL, Aberdeen papers, add mss 43230, fols 50–1.

32 *Quarterly Review*, vol. 2 (May 1809), pp. 348–55.

33 *Agricultural Magazine*, vol. 10 (January–June 1812), pp. 149–50.

34 P. Geoghegan, *The Irish Act of Union: a study in high politics* (Dublin, 1999); A. Blackstock, 'The Union and the military, 1810–c.1830', in *Transactions of the Royal Historical Society*, 6th series, vol. 10 (2000), pp. 329–51.

35 Hoppen, 'An incorporating Union?', pp. 335–6; S. J. Connolly, 'Union government, 1812–13', in W. Vaughan (ed.), *New history of Ireland, vol. 5: Ireland under the Union, 1801–70* (Oxford, 1989), pp. 66–70.

36 Memorandum presented by Col. Robert Hull to the Board of Agriculture as a plan for cultivating the waste, unenclosed and unproductive land of Great Britain, January–April 1802, PRONI, Foster–Massereene papers, D207/31/27; A. Horner, 'Napoleon's Irish legacy: the Bogs Commissioners, 1809–14', *History Ireland*, vol. 13, no. 5 (September–October 2005), pp. 24–8.

37 Malcomson, *John Foster*, p. 275; Sir John Sinclair to Foster, 5 February 1808, PRONI, Foster–Massereene papers, D562/5834; Richardson to Greenough, 18 May 1809, UCL, Additional Greenough papers, 1424; *Cobbett's parliamentary debates*, first series, vol. 14, pp. 337–8; Horner, 'Napoleon's Irish legacy', pp. 24–8.

38 Horner, 'Napoleon's Irish legacy', p. 25; Richardson to Greenough, 18 May 1809, UCL, Additional Greenough papers, 1424.

39 Richardson, *Essay on the culture of salt marshes*, pp. 34, 36.

40 W. Richardson, *Essay on the peat bogs of Ireland with observations on the mode of reclaiming them, directed by a late act* (London, 1812), pp. 1–4.

41 Richardson to Greenough, 5 February 1811, UCL, Additional Greenough papers, 1453.

42 Richardson to Greenough, 20 June, 15 November 1809, UCL, Additional Greenough papers, 1426, 1429.

43 *First report of the commissioners appointed to enquire into the nature and extent of the several bogs in Ireland*, [365] HC 1810, pp. 4, 7; Richardson to Greenough, 11 April 1810, UCL, Additional Greenough papers, 1439.

44 *Hansard's parliamentary debates*, 13 June 1809, vol. 14, p. 1015.

45 R. L. Edgeworth to Foster, 19, 20 April, 4 May 1810, PRONI, Foster–Massereene papers, T2519/904, 913, 934; Horner, 'Napoleon's Irish legacy', p. 25.

46 Richardson to Greenough, 28 July 1811; UCL, Additional Greenough papers, 1465; Richardson, *Peat bogs of Ireland*, pp. 1, 9.
47 Richardson to Greenough, 15 November 1809, UCL, Additional Greenough papers, 1430.
48 W. Richardson, *Peat bogs of Ireland*, preface; H. Davy to B. H. McCarthy, 1 February 1811, cited in *Second report of the commissioners appointed to enquire in to the nature and extent of Irish bogs*, [96] 1810–11, p. 205.
49 Mitchison, 'The old Board of Agriculture', p. 67.
50 Richardson, *Peat bogs of Ireland*, pp. 10, 20–5, 30–1, 33, 38.
51 Richardson, *Peat bogs of Ireland*, preface; *Second report of the commissioners*, p. 205.
52 Reverend Dr Rennie to Bishop Percy, 3 September 1810, in Nichols (ed.), *Illustrations of the literary history of the eighteenth century*, vol. 8, pp. 422–3.
53 Edgeworth to Foster, 19, 20 April, 4 May 1810, PRONI, Foster–Massereene papers, T2519/904, 913, 934.
54 Richardson to Greenough, 3 December 1811, UCL, Additional Greenough papers, 1472.
55 Richardson, *Essay on the improvement of Dartmoor*, pp. 140–1.
56 *Third report of the commissioners to enquire into the nature and extent of Irish bogs*, HC 130 (1813–14), vol. 6, pp. 4–5; *Fourth report of the commissioners to enquire into the nature and extent of Irish bogs*, HC 131 (1813–14), vol. 6, appendix to report 4, p. 173.
57 Richardson to Saurin, 22 January 1816, Saurin to Peel, 15 March 1816, BL, Peel papers, add mss, 40211, fols 157–60, 161–7.
58 Magennis, 'Coal, corn and canals', pp. 72–86; W. A. McCutcheon, 'The transport revolution: canals and river navigations', in K. B. Nowlan (ed.), *Travel and transport in Ireland* (Dublin, 1973), pp. 64–70.
59 W. Richardson, *Memoir on the subjects of making the Lower Bann navigable and restraining the winter redundancies of Lough Neagh* (Belfast, 1811), p. 3; Maguire, *Living like a lord*, p. 81.
60 Richardson, *Making the Lower Bann navigable*, pp. 8–9; Hilton, *Mad, bad and dangerous people?*, p. 257.
61 Richardson, *Making the Lower Bann navigable*, pp. 4, 7–10, 31–8, 48–9.
62 Richardson, *Making the Lower Bann navigable*, pp. 43–8; McCutcheon, 'Transport revolution', pp. 71–2.
63 Richardson, *Observations on Mr Townsend's plan of a canal linking Lough Neagh with the sea* (Newry, 1815), p. 3.
64 Richardson, *Observations on Mr. Townsend's plan*, pp. 3–12.
65 Richardson, *Making the Lower Bann navigable*, pp. 4, 11–13, 43.
66 T. Barnard, 'A clerical library in county Cork, 1774', in M. Fanning and R. Gillespie (eds), *Print culture and intellectual life in Ireland, 1660–1941* (Dublin, 2006), pp. 25–6, 40.
67 T. Barnard, *Improving Ireland: projects, prophets and profiteers, 1641–1786* (Dublin, 2008), p. 13.
68 Richardson to Greenough, 3 January 1808, UCL, Additional Greenough papers, 1407; Blackstock, *Ascendancy army*, p. 235; Senior, *Orangeism* pp. 46–7, 58; Blackstock, *Loyalism*, p. 134; Whelan, *Tree of Liberty*, p. 118.
69 Anketell, *A letter to Rev. Henry Conwell D.D.*
70 J. Kelly, 'Interdenominational relations and religious toleration in late eighteenth-century Ireland: the "paper war" of 1786–88', *Eighteenth-Century*

Ireland, vol. 3 (1988), pp. 63–4; H. Fenning, 'Dublin imprints of Catholic interest, 1783–89', *Collectanea Hibernica*, vols 44–45 (2002–03), p. 100; E. Lorraine de Montiluzan, 'The anti-Jacobin review after John Gifford by identified authors, 1807–21', *The Library*, vol. 6, no. 3 (September 2005), pp. 274–320.

71 Barnard, *Improving Ireland*, p. 113.

72 A. Blackstock, 'Politics and print: a case study', in R. Gillespie and A. Hadfield (eds), *History of the Irish book* (Oxford, 2006), p. 244.

73 For example, E. Ledwich, 'Discussion of a passage in the sixth Iliad of Homer', *Proceedings of the Royal Irish Academy*, vol. 4 (1790–92), pp. 3–11.

74 E. Ledwich, *Statistical account of the parish of Aghaboe* (Dublin, 1794).

75 R. Sweet, *Antiquaries: the discovery of the past in eighteenth-century Britain* (London and New York, 2004), pp. 11–13.

76 Ledwich, *Statistical account*, preface, pp. iii, iv.

77 J. Donaldson, *Agricultural biography* (London, 1854), pp. 107–8.

78 G. E. Fussell, *The English dairy farmer, 1500–1900* (London, 1966), p. 90.

79 Anonymous, *Memorandum containing accounts of experiments made in the cultivation of fiorin grass* (Edinburgh, 1820), p. 1.

80 *Caledonian Mercury*, 10 February 1820, 12 January 1822, 7 February 1824, 29 January 1825.

81 Richardson, *Essay on the improvement of Dartmoor*, p. 141.

82 *Statistical account of Scotland: account of 1834–45*, vol. 6, p. 611, and vol. 1, p. 360.

83 W. Jacob, *Observations arising from the benefits arising from the cultivation of poor soils by the application of pauper labour* (Lindfield, 1828); *Dublin Penny Journal*, vol. 1 (18 August 1832), pp. 59–60.

84 J. T. Mackay, 'Catalogue of the indigenous plants of Ireland', *Transactions of the Royal Irish Academy*, vol. 11 (1825), pp. 103–98.

85 J. C. Louden, *An encyclopaedia of agriculture* (London, 1825), pp. 819–22.

86 P. Lawson, *The agriculturalist's manual being a familiar description of the agricultural plants cultivated in Europe including practical observations respecting those suited to the climate of Great Britain* (Edinburgh, London and Dublin, 1836), p. 133.

87 P. Lawson, *Treatise on the cultivated grasses and other herbage and forage plants* (Edinburgh and London, 1842), p. 15.

Select bibliography

Primary manuscript sources

Armagh County Museum
Colonel Blacker's journals

British Library (BL)
Aberdeen papers, add mss 43229–30
Arthur Young papers, add mss 35130
Hardwicke papers, add mss 35739, 35750
Peel papers, add mss 40211
Pelham papers, add mss 33104

Kent Archives Office
Pratt papers, U840

Linenhall Library, Belfast (LHL)
Joy papers
Minute book of the Belfast Literary Society

National Archives of Ireland (NAI)
Prerogative grant book, 1839
Rebellion papers, 620 series

National Archives of Scotland (NAS)
Douglas-Hamilton papers, NRAS2177/944, 1575, 1584
Hannay papers, GD214
Miller of Dalswinton papers, GD197

National Library of Ireland (NLI)
Bruce MSS, 20888, 20898
Lake papers, MS56

National Library of Scotland (NLS)
Advocates' papers

Public Record Office (Kew)
Home Office (HO) 100 series

Public Record Office of Northern Ireland (PRONI)
Abercorn papers, D623, T2541, T3472
Agar papers, T3719
Armagh diocesan papers, DIO/4, DIO/32
Downshire papers, D671
Drennan transcripts, T765
Earl of Antrim papers, D2977
Emigrant letter, T1815
Foster–Massereene papers, D207, D562, T2519
Gage papers, T1883
Heyland papers, D4085
Hezlett papers, D668
Lenox-Conyingham papers, D1449
Macartney papers, D572, D2225
Maxwell Given papers, D2096
Ordnance Survey maps, D51

Robinson Library Armagh
Bound volume of Richardson's essays

Royal Institution
Humphry Davy's journals, HD15

Southampton University Library
Wellington papers, WP1

Suffolk County Record Office
Dr Richardson's letter to Brigadier-Major Kater, HA231

Tomas O'Fiaich Library, Armagh
Henry Conwell papers

Trinity College Dublin (Special Collections)
Richardson pamphlets

Ulster Museum
John Templeton's journals

University College London (UCL) (Special Collections)
Greenough papers
Additional Greenough papers
Greenough's journal, 7/13 (1806)

University of Glasgow (Special Collections)
Patrick Miller of Dalswinton papers

Newspapers

Belfast News-Letter
Caledonian Mercury
Derby Mercury
Lancaster Gazette
Morning Chronicle
Morning Post
The Times

Periodicals

Agricultural Magazine
Belfast Monthly Magazine
Critical Review or Annals of Literature
Farmer's Magazine
Munster Farmer's Magazine
Newry Magazine or Literary and Political Register
Quarterly Review
Weekly Entertainer or Agreeable and Instructive Repository (1812)

Primary printed source collections

Agnew, Jean and Luddy, Maria (eds), *The Drennan–McTier letters, 1802–1819* (3 vols, Dublin, 1999).

Chambers, Neil (ed.), *The scientific correspondence of Sir Joseph Banks* (6 vols, London, 2007).

Dawson, Warren R., *The Banks letters* (London, 1958).

Historical sketch of Belfast Literary Society, 1801–1901 (Belfast, 1902).

Hutton, Arthur Wollaston (ed.), *Arthur Young's tour in Ireland, 1776–1779* (2 vols, London, 1892).

Lockhart, J. G., *The memoirs of Sir Walter Scott* (4 vols, Paris, 1838).

McEvoy, J., *Statistical survey of the county of Tyrone* (Dublin, 1802).

Memoirs of Richard Lovell Edgeworth begun by himself and concluded by his daughter Maria Edgeworth (2 vols, London, 1820).

Miller, David W., *Peep O'Day Boys and Defenders: selected documents on the County Armagh disturbances, 1784–96* (Belfast, 1990).

Nichols, John Boyer (ed.), *Illustrations of the literary remains of the eighteenth century* (9 vols, London, 1858).

Vane, Charles (ed.), *The memoirs and correspondence of Viscount Castlereagh* (4 vols, London, 1848–53).

Wellington, *The supplementary despatches, letters and memoranda of Arthur Wellesley, first Duke of Wellington* (5 vols, London, 1860).

Parliamentary papers

First report of the commissioners appointed to enquire into the nature and extent of the several bogs in Ireland, [365] HC (1810–11).

Second report of the commissioners appointed to enquire into the nature and extent of Irish bogs, [96] HC (1810–11).

Third report of the commissioners to enquire into the nature and extent of Irish bogs, [130] HC (1813–14), vol. 6.

Fourth report of the commissioners to enquire into the nature and extent of Irish bogs, [131] HC (1813–14), vol. 6.

Report of the Select Committee appointed to enquire into the corn trade of the United Kingdom, [184] HC (1813), vol. 3.

The parliamentary debates from the year 1803 to the present time (cited as *Hansard's parliamentary debates*).

Primary printed sources

Ainslie, Robert, *Memoir on the fiorin grass* (Kirkudbright, 1811).

Antekell, John, *Strictures upon Paine's Age of Reason* (Dublin, 1796).

Anketell, John, *A letter to Rev. Henry Conwell D.D. popish priest of Drumglass and Killyman relative to a pamphlet lately published by Richard Wilson Esq.* (Dublin, 1807).

Bronson, E. (ed.), *Select reviews of literature* (Philadelphia, 1812).

Bruce, William, *The Christian soldier* (Belfast, 1803).

de Conynck, Frederick, *Fiorin Gräs eller Agrostis Stolonifera* (Copenhagen, 1816).

de Conynck, Frederick, *Afhandling om fiorin gräsets odling; författad på begäran af och till underwisning för hans kejserliga höghet, ärkehertig Johan af Österrike* (Stockholm, 1818).

Drewsen, Johannes Christian, *Om Krybehvene, Agrostis stolonifera (engelsk: Fiorin)* (Risbenhavn, 1818).

Dubordieu, J. (ed.), *A statistical survey of the county of Antrim* (2 vols, Dublin, 1812).

Edmundston, A., *A view of the ancient and present state of the Zetland Islands* (Edinburgh, 1809).

Farish, John, *Treatise on fiorin grass with a short description of its nature and properties, together with the soils and manures, best adapted to its culture* (Dumfries, 1810).

Henderson, John, *A general view of the agriculture in the county of Caithness* (London, 1815).

Musgrave, Sir Richard, *Memoirs of the different rebellions in Ireland* (Dublin, 1801).

Parnell, W., *An historical apology for the Irish Catholics* (Dublin, 1807, London, 1808).

Pictet, Marc-Auguste, *Voyage de trois mois, en Angleterre, en Écosse, et en Irelande* (Geneva, 1802).

Preston, Richard, *An address to the fund-holder, the mechanic, and the poor on the subject of the Corn Laws* (London, 1815).

Quayle, T., *General view of the agriculture of the Isle of Man* (London, 1812).

Richardson, W., *History of the origins of the Irish yeomanry* (Dublin, 1801).

——, 'Observations on the basaltic coast of Ireland in a letter transmitted from William Richardson DD by the Hon. George Knox MP', *Journal of Natural Philosophy, Chemistry and the Arts*, vol. 5 (1802), pp. 321–5.

——, 'On basalts', *Philosophical Magazine*, vol. 13 (1802), pp. 129–36.
——, 'An account of the whynn dykes in the neighbourhood of the Giant's Causeway, Ballycastle and Belfast in a letter to the Bishop of Dromore', *Transactions of the Royal Irish Academy*, vol. 9 (1803), pp. 21–43.
——, 'Inquiry into the consistency of Dr Hutton's theory of the earth with the arrangement of the strata and other phenomena on the basaltic coast of Antrim', *Transactions of the Royal Irish Academy*, vol. 9 (1803), pp. 429–87.
——, *Review of two memoirs in the Transactions of the Royal Irish Academy* (Dublin, 1805).
——, *Inquiry into the origin of the opinion, that basalt is a volcanic production, with the motives that induced its author [M. Desmerest] to publish it … with the facts and arguments adduced … in its support; and also the arguments and facts, from the Giant's Causeway and its vicinity, by which this opinion is proved unfounded* (Dublin, 1805).
——, 'Remarks on the basaltic coast of Antrim', *Transactions of the Royal Society of Edinburgh*, vol. 5 (1805), pp. 15–20.
——, 'On the volcanic theory part 1 Examination of Mr. Desmarest's memoir, in Transactions of the Academy of Sciences (1771); On the volcanic theory part 2 Examination of the facts and opinions, given by different advocates for the volcanic origin of basalt, who followed Mr. Desmarest: to wit, Mr. Faujas de St. Frond, Mr. Dolomieu, Mr. Whitehurst, Bishop Troll, Abbe Spalanzani, and Dr. Hamilton. Part 3 Arguments against the volcanic origin of basalt, derived from its arrangement in the county of Antrim, and from other facts observed in that county, communicated by the Rt. Reverend, the Lord Bishop of Dromore', *Transactions of the Royal Irish Academy*, vol. 10 (1806), pp. 35–108.
——, *An elementary treatise on the indigenous grasses of Ireland, with a selection of those which promise to be the most useful. Addressed to his agricultural friends* (Dublin, 1806).
——, *An essay on the improvement of the great flow bogs of Ireland particularly the Bog of Allen and the Montiaghs of the north … in a letter addressed to the Grand Juries of Antrim, Armagh and Tyrone* (Dublin, 1807).
——, *Memoir on fiorin grass to the Belfast Literary Society 1 March 1808* (Belfast, 1808).
——, *The reviewer reviewed; or, observations on a review of Dr Richardson's memoir on Irish fiorin grass as it was published in the Transactions of the Belfast Literary Society* (Belfast, 1808).
——, *Comment and notes on the preceding memoir as published by the Board of Agriculture* (Coleraine, 1808).
——, 'A letter on the alterations that have taken place in the structure of rocks, on the surface of the basaltic country in the counties of Derry and Antrim, from W. Richardson to H. Davy' [from the *Philosophical Transactions* of the Royal Society, vol. 118 (1808), pp. 187–222.
——, *Plan for reclaiming the Bog of Allen and the other great morasses in Ireland addressed to the Right Hon. Earl of Rosse* (Dublin, 1809).
——, 'Letter from W. Richardson, late fellow of Trinity College, Dublin to Thomas Allen Esq containing facts and observations relative to the variety of the Irish grass called FIORIN', *Farmer's Magazine*, vol. 40 (December 1809), pp. 503–10.
——, *Letter to the Rt. Hon. Isaac Corry, containing an epitome of some of the most curious and important properties of Irish fiorin, or fyoreen grass, with proofs that the facts by which they have been established, are fairly stated, and that the author is not mad* (Belfast, 1809).

——, *Essay on fiorin to the practical farmer* (London, 1810).
——, *Essay on fiorin grass shewing the circumstances under which it may be found in all parts of England* (London, 1810).
——, *Essay on the culture of salt marshes: the advantages of consigning them to fiorin grass alone, with the improvement they will receive from salt water irrigation* (Belfast, 1810).
——, *Letter to the most noble the Marquis of Hertford on fiorin grass containing the necessary directions for its culture; the periods and modes of laying it down....* (London, 1810).
——, *Letter to the most noble the Marquis of Hertford on fiorin grass* (London, 1810).
——, *Letter on the subject of reclaiming and improving the waste parts of the Highlands of Scotland ... by introducing the culture of fiorin grass* (London, 1810).
——, *Letter on irrigation to Right Hon. Isaac Corry* (Belfast, 1810).
——, *Epitome of a letter on irrigation now in the press, addressed to the Rt. Hon. Isaac Corry* (London, 1810).
——, *Memoir on the subjects of making the Lower Bann navigable and restraining the winter redundancy of Lough Neagh* (Belfast, 1811).
——, 'Memoir on useful grasses by W. Richardson, D.D. communicated by Dr Davenport read on Monday March 6 1809', *Transactions of the Royal Irish Academy*, vol. 11 (1810), pp. 87–119.
——, *Letter to the curious traveller, who shall visit the Giant's Causeway* (Coleraine, 1811).
——, *The utility of fiorin grass: a prize essay addressed to the Farming Society of Ireland, on their proposing to give a gold medal 'to the person who shall state to the Society the result of an experiment which shall be the most satisfactory in ascertaining the merit and utility of fiorin grass etc'* (London, 1811).
——, *Observations on the occasional scarcities, and poor laws in England with a remedy suggested by which a return of the former may be prevented and the pressure of the latter greatly abated; with a discussion of Mr Malthus's observations upon Ireland* (London, 1811).
——, 'On the strata of mountains', *Philosophical Magazine*, vol. 37 (1811), pp. 367–9.
——, *Memoir on the cultivation of fiorin grass in a letter addressed to the Bath and West of England Agricultural Society* (Bath, 1812).
——, *The correspondence between the Rev. Dr. Richardson and Mr. Tanner on fiorin grass* (London, 1812).
——, *Essay on the peat bogs of Ireland with observations on the mode of reclaiming them, directed by a late act* (London, 1812).
——, *Observations on a report of the agriculture and livestock of the county of Wicklow* (Dublin, 1812).
——, *Essay on the improvement of Dartmoor Forest, and on the means of so much increasing our grain crops as to make future importation unnecessary; both to be attained by the aid of fiorin grass*, printed in *Letters and papers on agricultural planting selected from the correspondences of the Bath and West of England Agricultural Society* (Bath and London, 1814), vol. 13, pp. 125–68.
——, *A new essay on fiorin grass including the history of its discovery, and an account of its valuable qualities and modes of culture* (London, 1814).
——, *Essay on the insufficiency of the Linnaean school to improve our knowledge in the natural history of the vegetable kingdom* (Newry, 1815).

——, 'Account of the peninsula of Portrush', *Newry Magazine or Literary and Political Register*, vol. 1 (1815), pp. 250–5.
——, *Directions for the cultivation of fiorin grass in the island of Newfoundland* (Newry, 1815).
——, *Observations on Mr Townsend's plan of a canal between Lough Neagh and the sea* (Newry, 1815).
——, *A second letter on agriculture as a science* (Newry, 1816).
——, 'Simple measures by which the recurrence of famines may be prevented and the pressure of the poor laws greatly abated by a slight and partial change in our common agricultural practice', *Pamphleteer*, vol. 7 (1816), pp. 156–214.
——, *A description of the basaltic coast from Magilligan to Glenarm* (Newry, 1817).
——, *Observations on memoirs published by the Geological Society, respecting the north-eastern counties of Ireland* (Newry, 1817).
——, *Letter on the improvement of grassy mountains, detailing the measures by which they may be made to maintain through winter the whole stock that grazed upon them in summer* (Newry, 1817).
——, *An essay on agriculture; containing an introduction in which the science of agriculture is pointed out, by a careful attention to the works of nature; also the means of rendering barren soils luxuriantly productive, at a very moderate expense, and of beneficially employing the industrious and unoccupied poor. To which is added a memoir drawn up at the express wish of his Imperial Highness the Arch-duke John of Austria, on the nature and qualities of fiorin grass, with practical remarks on its abundant properties, and the best means of cultivating that extraordinary vegetable* (London, 1818).
——, *Improvement of heathy and peaty mountains* (Belfast, 1818).
Ricky, Walter, *The extraordinary conduct of the Rev. Dr. Richardson, rector of the parish of Clonfeacle … towards the curate of the same parish* (Dublin, 1807).
——, *Dr Richardson's calumny refuted* (Dungannon, 1808).
Sinclair, Sir John, *Address to the Board of Agriculture* (London, 1808).
Singer, Dr, *General view of the agriculture … in the county of Dumfries* (Edinburgh, 1812).
Smith, J., *Essay on indigenous grasses* (Dublin, 1808).
Trimmer, J. K., *A brief enquiry into the state of agriculture in the southern part of Ireland* (London, 1809).
Wade, W., *Sketch of lectures on meadow and pasture grasses* (Dublin, 1808).
Wakefield, Edward, *Account of Ireland statistical and political* (2 vols, London, 1812).
Wilson, R., *A narrative of the various murders and robberies committed … upon the Roman Catholics by a banditti describing themselves Orangemen* (Dublin, 1808).

Secondary sources

Adelman, Juliana, *Communities of science in nineteenth-century Ireland* (London, 2009).
Agnew, J., *Belfast merchant families in the seventeenth century* (Dublin, 1996).
Armstrong, P., *The English parson-naturalist: a companionship between religion and science* (Leominster, 2000).
Bardon, J., *Belfast: an illustrated history* (Belfast, 1982).
——, *A history of Ulster* (Belfast, 1992).

Barnard, T., 'Improving clergymen, 1660–1760', in A. Ford, J. McGuire and K. Milne (eds), *As by law established: the Church of Ireland since the Reformation* (Dublin, 1995), pp. 136–51.
——, 'The Dublin Society and other improving societies, 1731–85', in J. Kelly and M. J. Powell (eds), *Clubs and societies in eighteenth-century Ireland* (Dublin, 2010), pp. 53–88.
——, *Improving Ireland: projects, prophets and profiteers, 1641–1786* (Dublin, 2008).
——, *The kingdom of Ireland, 1641–1760* (Basingstoke, 2004).
——, *A new anatomy of Ireland: the Irish Protestants, 1649–1770* (New Haven and London, 2003).
Bartlett, T., *The fall and rise of the Irish nation: the Catholic question, 1690–1830* (Dublin, 1992).
Bayles, R., 'Understanding local science: the Belfast Natural History Society in the mid-nineteenth century', in D. Altis and C. Mollan (eds), *Science and Irish culture* (Dublin, 2004), vol. 1, pp. 139–69.
Bebbington, David W., 'Science and evangelical theology in Britain from Wesley to Orr', in David N. Livingstone, D. G. Hart and Mark A. Noll (eds), *Evangelicals and science in historical perspective* (Oxford, 1999), pp. 120–41.
Benn, G., *A history of the town of Belfast* (2 vols, London, 1880).
Betham-Edwards, M. (ed.), *The autobiography of Arthur Young* (London, 1898).
Bew, John (ed.), *Belfast politics: thoughts on the British constitution by William Bruce and Henry Joy* (reprint Dublin, 2005).
——, *The glory of being Britons: civic unionism and nineteenth-century Belfast* (Dublin, 2009).
Blackstock, A., *An ascendancy army: the Irish yeomanry, 1796–1834* (Dublin, 1998).
——, *Double traitors: the Belfast volunteers and yeomanry, 1778–1828* (Belfast, 2000).
——, 'The invincible mass: loyal crowds in mid-Ulster, 1795–96', in P. J. Jupp and E. Magennis (eds), *Crowds in Ireland, c.1720–1920* (Basingstoke and New York, 2000), pp. 83–114.
——, 'The Union and the military, 1810–c.1830', *Transactions of the Royal Historical Society*, 6th series, vol. 10 (2000), pp. 329–51.
——, 'The rector and the rebel', in S. Wichert (ed.), *From the United Irishmen to twentieth-century unionism: a festschrift for A. T. Q. Stewart* (Dublin, 2004).
——, *Loyalism in Ireland, 1789–1829* (Woodbridge, 2007).
——, 'Loyal clubs and societies in Ulster, 1770–1800', in J. Kelly and M. J. Powell (eds), *Clubs and societies in eighteenth-century Ireland* (Dublin, 2010), pp. 447–65.
——, 'Loyalist associational culture and civic liberty in Belfast, 1793–1835', in Jennifer Kelly and V. Comerford (eds), *Associational culture in Ireland and abroad* (Dublin, 2010), pp. 47–66.
Borsay, P., 'The culture of improvement', in P. Langford (ed.), *The eighteenth century* (Oxford, 2002), pp. 183–212.
Bright, K., *The Royal Dublin Society, 1815–1845* (Dublin, 2004).
Brown, M., 'Configuring the Irish Enlightenment: reading the *Transactions of the Royal Irish Academy*', in J. Kelly and M. J. Powell (eds), *Clubs and societies in eighteenth-century Ireland* (Dublin, 2010), pp. 163–80.
Clark, Peter, *British clubs and societies, 1580–1800: the origins of an associational world* (Oxford, 2000).
Collini, S., Winch, D., and Burrow, J. (eds), *That noble science of politics: a study in nineteenth-century intellectual history* (Cambridge, 1983).

Connolly, S. J., 'Union government, 1812–13', in W. Vaughan (ed.), *New history of Ireland, vol. 5: Ireland under the Union, 1801–70* (Oxford, 1989).

——, *Religion, law and power: the making of Protestant Ireland, 1660–1760* (Oxford, 1992).

——, *Divided kingdom: Ireland, 1630–1800* (Oxford, 2008).

Cookson, J., *The British armed nation* (Oxford, 1997).

Cousins, G., *The defenders: a history of the British volunteers* (London, 1968).

Curtin, N., *The United Irishmen: popular politics in Ulster and Dublin, 1791–98* (Oxford, 1994).

Dickson, D., *Old world colony: Cork and south Munster, 1630–1830* (Cork, 2005).

Digby, A., 'Malthus and reform of the poor law', in J. Dupaquier and E. Grebenik (eds), *Malthus past and present* (London and New York, 1983), pp. 97–109.

Dixon, Thomas, 'Introduction', in T. Dixon, G. Cantor and S. Pumfrey (eds), *Science and religion: new historical perspectives* (Cambridge, 2010), pp. 1–19.

Eastwood, David,. 'E. P. Thompson, Britain and the French Revolution', *History Workshop Journal*, vol. 39 (spring 1995), pp. 79–101.

——, *Government and community in the English provinces, 1700–1870* (Basingstoke, 1997).

Elliott, Marianne, *Partners in Revolution: the United Irishmen and France* (New Haven and New York, 1983).

——, *Wolfe Tone: prophet of Irish independence* (New Haven and London, 1989).

Emerson, Roger L., 'The Philosophical Society of Edinburgh, 1768–83', *British Journal of the History of Science*, vol. 18 (1985), pp. 255–303.

Fissell, Mary and Cooter, Roger, 'Exploring natural knowledge: science and the popular', in Roy Porter (ed.), *The Cambridge history of science, vol. 4: eighteenth-century science* (Cambridge, 2003), pp. 129–59.

Foster, John Wilson (ed.), *Nature in Ireland: a scientific and cultural history* (Dublin, 1997).

Fussell, G. E., *The English dairy farmer, 1500–1900* (London, 1966).

Gambles, A., *Protection and politics: conservative economic discourse, 1815–1852* (Woodbridge, 1999).

Geoghegan, P., *The Irish Act of Union: a study in high politics* (Dublin, 1999).

Golinski, J., *Science as public culture: chemistry and Enlightenment in Britain, 1760–1820* (Cambridge, 1992).

Grand Orange Lodge of Ireland, Education Committee, *The formation of the Orange Order, 1795–1798* (Belfast, 1994).

Gray, Peter, *The making of the Irish poor law, 1815–43* (Manchester, 2009).

Herbert, Sandra, *Charles Darwin, geologist* (London, 2005).

Herries Davies, Gordon L., 'Geology in Ireland before 1812: a bibliographical outline', *Western Naturalist*, vol. 11 (1978), pp. 79–99.

——, 'The history of the earth sciences', in D. G. Smith (ed.), *Cambridge encyclopaedia of earth sciences* (Cambridge, 1981), pp. 11–23.

——, 'Astronomy, geology and mineralogy', in T. O'Raifeartaigh (ed.), *The Royal Irish Academy: a bicentennial history* (Dublin, 1985), pp. 242–73.

Hill, Jacqueline, *From patriots to unionists: Dublin civic politics and Irish Protestant patriotism, 1660–1840* (Oxford, 1997).

Hilton, Boyd, *A mad, bad and dangerous people? England, 1783–1846* (Oxford, 2006).

Holmes, Andrew R., *The shaping of Ulster Presbyterian belief and practice, 1770–1840* (Oxford, 2006).

——, 'Presbyterians and science in the north of Ireland before 1874', *British Journal of the History of Science*, vol. 41, no. 4 (December 2008), pp. 540–60.

Holmes, R., *The age of wonder* (London, 2008).

Hoppen, Theo, 'An incorporating Union? British politicians and Ireland, 1800–1830', *EHR*, vol. 123, no. 501 (April 2008), pp. 328–50.

Horner, A., 'Napoleon's Irish legacy: the Bog Commissioners, 1809–14', *History Ireland*, vol. 13, no. 5 (September–October 2005), pp. 24–8.

Inkster, I., 'The development of a scientific community in Sheffield, 1790–1850: a network of people and interests', *Transactions of the Hunter Archaeological Society*, vol. 10 (1973), pp. 99–131.

Jackson, P. N. Wyse, 'Tumultuous times: geology in Ireland and the debate on the nature of basalt and other rocks of north-east Ireland between 1740 and 1816', in P. N. Wyse Jackson (ed.), *Science and engineering in Ireland in 1798* (Dublin, 2000), pp. 35–50.

James, Kevin J., *Handloom weavers in the Ulster linen industry, 1815–1914* (Dublin, 2007).

James, P., *Population Malthus* (London, 1979).

Jamieson, J., *The history of the Royal Belfast Academical Institution, 1810–1960* (Belfast, 1960).

Jupp, P. J., *British politics on the eve of reform* (Basingstoke, 1998).

——, *The governing of Britain, 1688–1848* (Abingdon and New York, 2006).

Kelly, James, *Prelude to Union: Anglo-Irish politics in the 1780s* (Cork, 1992).

——, *Sir Edward Newenham MP 1793–1814: defender of the Protestant constitution* (Dublin, 2004).

——, *Sir Richard Musgrave, 1746–1818, ultra-Protestant ideologue* (Dublin, 2009).

Kelly, P., 'The politics of political economy in mid-eighteenth-century Ireland', in S. J. Connolly (ed.), *Political ideas in eighteenth-century Ireland* (Dublin, 2000), pp. 105–29.

Keogh, D., *The French disease: the Catholic Church and radicalism in Ireland, 1790–1800* (Dublin, 1993).

Killen, J., *A history of the Linenhall Library, 1788–1988* (Belfast, 1990).

——, 'The reading habits of a Georgian gentlemen, John Templeton, and the book collections of the Belfast Society for Promoting Knowledge', in B. Cunningham and M. Kennedy (eds), *The experience of reading: Irish historical perspectives* (Dublin, 1999), pp. 99–108.

Larmour, P. and McBride, S., 'Buildings and faith: church building from medieval to modern', in R. Gillespie and W. G. Neely (eds), *The laity and the Church of Ireland, 1000–2000* (Dublin, 2002), pp. 304–50.

Lecky, W. E. H, *Ireland in the eighteenth century* (Cabinet edition, 4 vols, London, 1913).

Livesay, J., 'The Dublin Society in eighteenth-century Irish political thought', *Historical Journal*, vol. 47 (2004), pp. 615–40.

Magennis, E., 'A "Presbyterian insurrection"? Reconsidering the Hearts of Oak disturbances of July 1763', *Irish Historical Studies*, vol. 31, no. 122 (November 1998), pp. 165–87.

——, 'Coal, corn and canals: parliament and the dispersal of public moneys, 1695–1772', in D. Hayton (ed.), *The Irish parliament in the eighteenth-century* (Edinburgh, 2000), pp. 71–86.

——, 'A land of milk and honey: the Physico-Historical Society, improvement and the surveys of mid-eighteenth-century Ireland', *Proceedings of the Royal Irish Academy*, vol. 102C, no. 6 (2002), pp. 199–217.
——, 'Clubs and societies in eighteenth-century Belfast', in J. Kelly and M. J. Powell (eds), *Clubs and societies in eighteenth-century Ireland* (Dublin, 2010), pp. 466–84.
Maguire, W. A., *Living like a lord: the second Marquis of Donegall, 1769–1844* (Belfast, 1984).
Malcomson, A. P. W., *John Foster: the politics of the Anglo-Irish ascendancy* (Oxford, 1978).
——, *Archbishop Charles Agar: churchmanship and politics in Ireland, 1760–1810* (Dublin, 2002).
——, *John Foster: the politics of improvement and prosperity* (Dublin, 2011).
——, 'A lost natural leader: John James Hamilton first Marquis of Abercorn', *Proceedings of the Royal Irish Academy*, vol. 88, C, no. 4 (1988), pp. 61–86.
——, 'Introduction to the Abercorn papers in PRONI', PRONI, D623.
Mansergh, D., *Grattan's failure: parliamentary opposition and the people in Ireland, 1779–1800* (Dublin, 2005).
Maxwell, C., *A history of Trinity College Dublin, 1591–1892* (Dublin, 1946).
McAnally, Sir Henry, *The Irish militia, 1793–1816* (Dublin and London, 1949).
McBride, I., *The siege of Derry in Ulster Protestant mythology* (Dublin, 1997).
——, *Scripture politics: Ulster Presbyterians and Irish radicalism in the late eighteenth century* (Oxford, 1998).
——, *Eighteenth-century Ireland* (Dublin, 2009).
McCutcheon, W. A., 'The transport revolution: canals and river navigations', in Kevin B. Nowlan (ed.), *Travel and transport in Ireland* (Dublin, 1973), pp. 64–81.
McDowell, R. B., *Ireland in the age of imperialism and revolution, 1760–1801* (Oxford, 1979).
—— and Webb, D. A., *Trinity College Dublin, 1592–1952: an academic history* (Cambridge, 1982).
McGleenon, C. F., *A very independent county: parliamentary elections and politics in County Armagh, 1750–1800* (Belfast, 2011).
Miller, David Philip, 'The usefulness of natural philosophy: the Royal Society and the culture of practical utility in the later eighteenth century', *British Journal of the History of Science*, vol. 32 (1999), pp. 185–201.
Mirala, P., *Freemasonry in Ulster, 1733–1813* (Dublin, 2007).
Mitchinson, R., 'The old Board of Agriculture, 1793–1822', *English Historical Review*, vol. 74 (1959), pp. 41–69.
——, *Agricultural Sir John: the life of Sir John Sinclair of Ulster* (London, 1962).
Mullin, T. H., *Coleraine in Georgian times* (Belfast, 1997).
Nelson, I., *The Irish militia, 1793–1802: Ireland's forgotten army* (Dublin, 2007).
O'Gorman, Frank, 'English loyalism re-visited', in A. Blackstock and E. Magennis (eds), *Politics and political culture in Britain and Ireland* (Belfast, 2007), pp. 223–41.
O'Riordan, M., 'The native Ulster mentalité as revealed in Gaelic sources, 1600–1650', in B. Mac Cuarta (ed.), *Ulster 1641: aspects of the rising* (Belfast, 1997), pp. 61–91.
Patten, E., *Samuel Ferguson and the culture of nineteenth-century Ireland* (Dublin, 2004).
Phillipson, N., *Adam Smith, an enlightened life* (London, 2010).
Porter, R., *The making of geology: earth science in Britain, 1660–1815* (Cambridge, 1977).

——, 'Gentlemen and geology: the emergence of a scientific career, 1660–1920', *Historical Journal*, vol. 21, no. 4 (1978), pp. 809–36.

Purver, M., *The Royal Society: concept and creation* (London, 1967).

Rappaport, R., 'The earth sciences', in R. Porter (ed.), *Cambridge history of science, vol. 4: eighteenth-century science* (Cambridge, 2003), pp. 412–35.

Rashid, Salim, 'Political economy and geology in the early nineteenth century: similarities and contrasts', *History of Political Economy*, vol. 13 (1981), pp. 726–44.

——, 'Dugald Stewart, "Baconian" methodology, and political economy', *Journal of the History of Ideas*, vol. 46, no. 2 (1985), pp. 245–57.

Rudwick, Martin J. S., *Worlds before Adam: the reconstruction of geology in the age of reform* (Chicago and London, 2008).

Semmell, B. (ed.), *Occasional papers of T. R. Malthus* (New York, 1963).

Senior, H., *Orangeism in Ireland and Britain, 1795–1836* (London and Toronto, 1966).

Shapin, S., 'The Pottery Philosophical Society, 1819–1835: an examination of the cultural uses of provincial science', *Social Studies*, vol. 2 (1972), pp. 311–36.

——, 'Nibbling at the teats of science: Edinburgh and the diffusion of science in the 1830s', in I. Inkster and J. Morell (eds), *Metropolis and province: science in British culture 1780–1850* (London, 1983), pp. 151–72.

Simms, H., 'Violence in County Armagh', in B. Mac Cuarta (ed.), *Ulster 1641: aspects of the Rising* (Belfast, 1997), pp. 123–38.

Sneddon, A., 'Bishop Francis Hutchinson (1660–1739): a case study in the culture of eighteenth-century improvement', *Irish Historical Studies*, vol. 35, no. 139 (May 2007), pp. 289–310.

——, 'Legislating for economic development: Irish fisheries as a case study of the limits of "improvement"', in D. W. Hayton, J. Kelly and J. Bergin (eds), *The eighteenth-century composite state: representative institutions in Ireland and Europe, 1689–1800* (Basingstoke, 2010).

Stewart, A. T. Q., *A deeper silence: the hidden origins of the United Irishmen* (London and Boston, 1993).

——, *The summer soldiers: the 1798 rebellion in Antrim and Down* (Belfast, 1995).

Sweet, R., *Antiquaries: the discovery of the past in eighteenth-century Britain* (London and New York, 2004).

Thackray, A., 'Natural knowledge in cultural context: the Manchester model', *American History Review*, vol. 79, no. 3 (1974), pp. 672–709.

Uglow, J., *The lunar men* (London, 2002).

Whelan, K., *The tree of liberty: radicalism, Catholicism and the construction of Irish identity* (Cork, 1996).

White, T. de Vere, *The story of the Royal Dublin Society* (Tralee, 1955).

Wilson, James, 'Orangeism in 1798', in T. Bartlett, D. Dickson, D. Keogh and K. Whelan (eds), *1798: a bicentenary perspective* (Dublin, 2003), pp. 345–62.

Winch, D., 'Higher maximum: happiness versus wealth in Malthus and Ricardo', in S. Collini, D. Winch and J. Burrow (eds), *That noble science of politics: a study in nineteenth-century political history* (Cambridge, 1983), pp. 63–77.

Withers, Charles W. J., 'Improvement and enlightenment: agriculture and natural history in the work of the Rev. Dr. John Walker (1731–1803)', in P. Jones (ed.), *Philosophy and science in the Scottish Enlightenment* (Edinburgh, 1988), pp. 102–16.

Woodward, Horace B., *A history of the Geological Society of London* (London, 1907).

Young, A. I., *Three hundred years in Innishowen* (Belfast, 1919).
Young, Davis A., *Mind over magma: the story of igneous petrology* (Princeton, 2003).
Young, W. R., *The fighters of Derry* (London, 1932).

Reference books

Burtchaell, G. D. and Sadlier, T. U. (eds), *Alumni Dublinenses, 1593–1860* (Dublin, 1935).
Day, A. and P. McWilliams, P. (eds), *Ordnance Survey memoirs of Ireland, volume 1: Parishes of County Armagh, 1835–8* (Belfast, 1990).
Dictionary of Irish biography (9 vols, Cambridge, 2009).
Donaldson, J., *Agricultural biography* (London, 1854).
Fleming, W. E. C., *Armagh clergy, 1800–2000* (Dundalk, 2000).
George, M. D., *Catalogue of political and personal satires* (London, 1952).
Golden, J., *A list of the papers and correspondence of George Bellas Greenough* (London, 1981).
Johnston-Liik, E. M. (ed.), *History of the Irish parliament* (6 vols, Belfast, 2002).
Lewis, S., *Counties Fermanagh and Tyrone: a topographical dictionary* (London, 1837, new edition, Belfast, 2004).
New Oxford dictionary of national biography (60 vols, Oxford, 2004).
Thorne, R. G. (ed.), *The House of Commons, 1790–1820* (5 vols, London, 1986).

Index